JAMES CLERK MAXWELL

1831–1931

JAMES CLERK MAXWELL

JAMES CLERK MAXWELL

A Commemoration Volume

1831–1931

ESSAYS

by

Sir J. J. Thomson
Max Planck
Albert Einstein
Sir Joseph Larmor
Sir James Jeans
William Garnett
Sir Ambrose Fleming
Sir Oliver Lodge
Sir R. T. Glazebrook
Sir Horace Lamb

CAMBRIDGE

At the University Press

1931

CAMBRIDGE UNIVERSITY PRESS
Cambridge, New York, Melbourne, Madrid, Cape Town,
Singapore, São Paulo, Delhi, Tokyo, Mexico City

Cambridge University Press
The Edinburgh Building, Cambridge CB2 8RU, UK

Published in the United States of America by Cambridge University Press, New York

www.cambridge.org
Information on this title: www.cambridge.org/9781107670952

First published 1931
First paperback edition 2011

A catalogue record for this publication is available from the British Library

ISBN 978-1-107-67095-2 Paperback

CONTENTS

PORTRAITS

JAMES CLERK MAXWELL

BY

Sir J. J. Thomson

WE are met to celebrate the centenary of one whose work has had a profound influence on the progress and conceptions of Physical Science; it has moreover been instrumental in harnessing the ether for the service of man and has thereby advanced civilization and increased the safety and happiness of mankind.

Maxwell came of a race, the Clerks of Penycuik in Midlothian, who for two centuries had been prominent in the social life of Scotland; each generation had been remarkable for the talents and accomplishments of some of its members; one of these, Will Clerk, was the intimate friend of Sir Walter Scott and the original of the Darsie Lattimer of *Redgauntlet*. As a race they were remarkable, like Maxwell himself, for strong individuality.

John Clerk Maxwell, Maxwell's father, had added the name of Maxwell to that of Clerk on inheriting the small estate of Middlebie in Dumfriesshire. His main characteristic according to Lewis Campbell was a persistent practical interest in all useful processes. He was called, like so many of his class, to the Scottish Bar but does not seem to have done much business. His interests were in mechanical contrivances, and his acme of felicity was to attend a meeting of the Royal Society of Edinburgh. He had ways of his own of doing most things, even in designing clothes for himself and his son, which led to disastrous results when the boy went to school. Maxwell's mother was Frances Cay, a member of a well-known Northumbrian family.

Maxwell was born in Edinburgh on June 13, 1831, but spent his infancy and early boyhood at Glenlair, a house built by his father shortly after his marriage. The child took great delight in the happenings of country life and seems to have had more than the usual share of childish inquisitiveness. His cousin Mrs Blackburne said that throughout his childhood his constant question was "What's the go of that?", "What does it do?" and if he was not satisfied with the answer he would ask "But what is the 'particular go' of it?" Besides asking questions he was very fond of making things such as baskets and seals covered with strange devices.

His mother died, when he was in his ninth year,

of the disease which killed him at the same age forty years later. After his mother's death his great delight seems to have been to go about with his father and "help" him when he was doing jobs on the estate. The relations between father and son were extraordinarily intimate. When he was at school his letters to his father were like letters to another schoolboy, and full of the quips and dry humour which were characteristic of him throughout his life.

When he was ten years old he went to the Edinburgh Academy, and was at first anything but a success. There were many reasons for this. He entered at the middle of the term; he had mixed very little with other boys and was naturally shy and awkward; he was not at all well prepared in school subjects and had a very strong Galloway accent, but worst of all he wore clothes designed by his father on what would now be called hygienic principles; he had a lace frill instead of a collar round his neck, a tunic instead of a coat, and square-toed shoes of a novel pattern, with a brass buckle. All these naturally called for vigorous protest, with the result that when he returned home the skirt of his tunic was missing and his frill was rumpled and torn. Things however slowly improved, Professor P. G. Tait who was his junior by one year at the Academy says: "At school he was at first regarded as shy and rather dull. He made no friendships and spent his occasional holidays in reading old ballads, drawing curious diagrams and making rude mechanical models. This absorption in such pursuits, totally unintelligible to his schoolfellows, who were then totally ignorant of mathematics, procured him a not very complimentary nickname.[1] About the middle of his school career however he surprised his companions by suddenly becoming one of the most brilliant among them, gaining prizes and sometimes the highest prizes for scholarship, mathematics, and English verse."

Before he was fifteen he had written a paper on a mechanical method of describing curves of a certain type, which was published in the *Proceedings* of the Royal Society of Edinburgh. It was found afterwards that in this method he had been anticipated by no less famous a mathematician than Descartes. Tait says that at the time the paper was written he had received no instruction in mathematics beyond a few books of Euclid and the merest elements of algebra.

[1] "Dafty."

It was early in his school career that he began to write verses, a practice which he kept up all his life, to the great delight of his friends. Many of these, including two famous in the domestic history of Trinity College, "John Alexander Frere, John" and the ode to the portrait of Cayley, are given in Lewis Campbell's life of Maxwell. The last he ever wrote, "A Paradoxical Ode" to Hermann Stoffkraft, written in 1878, was called forth by the publication of a book, *Paradoxical Philosophy*, by two of his intimate friends P. G. Tait and Balfour Stewart. It has some lines which are a remarkable anticipation of the speculations which are now so common about the destiny of matter and energy:

> Till in the twilight of the Gods,
> When earth and sun are frozen clods,
> When all its energy degraded,
> Matter to ether shall have faded;
> We, that is, all the work we've done
> As waves in ether shall for ever run
In ever widening spheres through heavens beyond the sun.

During his school-days we first hear of a game which he played throughout his life and which all his friends associate with him. One name for it is "diabolo", but it was usually called the "devil on two sticks". The devil consists of a double cone, the narrow part resting on a string, whose ends are attached to the ends of two sticks which are held in the hands of the player; by moving the ends of the sticks in opposite directions it is possible to give very considerable rotation to the devil; in fact it is a home-made gyroscope with all the paradoxical properties of that instrument. He attained great skill with it, and no doubt it led him to the construction of his dynamical top, by which he demonstrated in a striking way the properties of bodies in rotation. Another toy which attracted him in boyhood and to which later on he also gave a scientific application was the zoetrope or wheel of life. Long afterwards he used it to represent the way two circular vortex rings play at leap-frog with each other. This is I think the first application of the principle of the cinematograph to scientific purposes.

Maxwell spent six years at the Academy and then went for three years to the University of Edinburgh, and was allowed by Professor Forbes to use the lecture apparatus for his own experiments. He read widely, though not systematically, in Mathematical Physics and published two papers in the *Transactions* of the Royal Society of Edinburgh. On leaving Edinburgh in 1850 he proceeded to Cambridge, entering first at Peterhouse where Tait had preceded him, but migrating after one term to Trinity, the principal reason being that there was at that time a number of very able mathematicians at Peterhouse and the chance of a mathematician obtaining a Fellowship there seemed less than at Trinity. He brought to Cambridge a range of mathematical knowledge extraordinary for so young a man, but in a state of disorder which Tait said appalled his private tutor Hopkins. Early in his second year of residence he was elected to a Scholarship at Trinity. At that time, and indeed for long afterwards, the Scholars dined together at one table. This brought Maxwell into daily contact with the most intellectual set in the College, among whom were many who attained distinction in later life. These in spite of his shyness and some eccentricities recognized his exceptional powers. He was made one of the "Apostles", a club limited to 12 members, who in the opinion of the Club were the men of most outstanding ability in the University.

Dr Butler, who later became Master of Trinity, says: "When I came up to Trinity, James Clerk Maxwell was just beginning his second year. His position among us was unique. He was the one acknowledged man of genius among the Undergraduates". The impression of power which Maxwell produced on all he met was remarkable; it was often much more due to his personality than to what he said, for many found it difficult to follow him in his quick changes from one subject to another, his lively imagination started so many hares that before he had run one down he was off after another. I was told by Dr Butler that he remembered going for a walk with Maxwell without understanding one word of what he said though he talked the whole of the time, and yet, said Dr Butler, "I would not have missed it for anything". We have evidence of the charm of his conversation in the diary of a great friend of his, a Mathematical Scholar of his year, who writes: "Maxwell as usual showing himself acquainted with every subject on which the conversation turned. I never met a man like him. I do believe there is no subject on which he cannot talk and talk well too, displaying always the most curious and out of the way information". Another friend writes: "Among his friends he was the most genial and amusing of companions, the propounder of many a strange theory, the composer of many a poetic *jeu d'esprit*".

At the beginning of his second year at Cambridge he began his serious preparation for the Mathematical Tripos. Though he had read much more widely than his contemporaries it had been in a very desultory fashion and was not likely to show to advantage in any examination where speed was a factor of considerable importance. He set himself resolutely to remedy this defect. Although preparing for an examination is regarded as almost degrading by some who are keenly conscious of their own superiority, Maxwell in his correspondence at this time never expresses any irritation, though he followed loyally and closely the normal course for the best men. He became a pupil of Hopkins, a man of great scientific distinction, as well as the most successful mathematical teacher of his time, and who had had Stokes and William Thomson among his pupils, and was careful to do all the work Hopkins set. There are references in letters of this period to his breaking away from parties to go and do "old Hop's props". The term before the Tripos was spent in revising subjects read before. Many find this distasteful, but Maxwell writes: "If any one asks how I am getting on in Mathematics say that I am busy arranging everything so as to be able to express all distinctly so that examiners may be satisfied now and pupils edified hereafter. It is pleasant work and very strengthening but not nearly finished".

Besides reading with Hopkins he attended Stokes' lectures. He took the Mathematical Tripos in January 1854 and was Second Wrangler, Routh of Peterhouse, who became a most successful teacher of Mathematics with twenty-seven Senior Wranglers to his credit, and who also made original contributions to Mathematics of a very high order, being Senior. In the examination for the Smith's Prize, where the papers are confined to the higher subjects, the two were bracketed.

It is interesting that in the paper set by Stokes in this examination there was published for the first time the fundamentally important theorem known as Stokes' theorem, connecting a line with a surface integral; this was to prove of vital importance in the development of Maxwell's electric theory.

Hopkins is reported to have said that Maxwell was unquestionably the most extraordinary man he had met with in the whole course of his experience; that it appeared impossible for him to think wrongly on any physical subject, but that in analysis he was far more deficient. Maxwell's preference for geometrical methods over analytical ones is mentioned by many of his contemporaries. This is an example of his general method of thought, which was to proceed step by step, from one definite idea to another, until he had reached the goal, instead of getting there by means of symbols and equations without any visualization of the intermediate stages.

After his degree Maxwell stayed up at Trinity, taking pupils, working at his theory of Colour and reading Faraday's *Experimental Researches*. In 1855 he was elected to a Fellowship and took his share in the College work; at the end of the year he published his paper on "Faraday's Lines of Force". His father's health had now broken down and was giving him much anxiety, making him want to spend as much time at home as possible; this was one of the reasons why he became a candidate for the Professorship in Natural Philosophy at the Marischal College, Aberdeen. His father, however, died in 1856 before the appointment was announced. "The personal loss to him", says Mr Lewis Campbell, "was irreparable, their long daily companionship had been followed by a correspondence which was all but daily, by vacations spent together, and an uninterrupted interchange, whether present or absent, of thought and social interest, both light and grave."

Maxwell was elected to the Chair at Marischal College soon after his father's death, but did not begin work there until November; he returned to Cambridge where he remained until June doing College work and also teaching a class of working men. He also published a paper on "Geometrical Optics". He had, when reading the work of Smith and Cotes on this subject, been struck by the superiority of the old methods over those then prevalent in Cambridge. In his paper he introduces some of these methods and adds others. There is a characteristic freshness and elegance about Maxwell's papers on "Geometrical Optics" which make them very delightful reading.

In November 1856 he began work at Aberdeen. It is clear from his letters at this time that he devoted a great deal of thought and time to his classes and to the preparation of striking experiments for his lectures. He seems too to have kept a careful watch on his tongue, for in a letter to Lewis Campbell after he had been two months at Aberdeen he writes: "One thing I am thankful for though perhaps you will not believe it; up to the present time I have not even been tempted to

CLERK MAXWELL AS A YOUNG MAN

mystify anyone", and again: "No jokes of any kind are understood here, I have not made one for two months and if I feel one coming I shall bite my tongue". His chief scientific work during the first two years at Aberdeen was his Adams Prize Essay on "Saturn's Rings". This prize is awarded for the best solution of some problem of great scientific importance suggested by the electors. In this case the subject selected was the motion of Saturn's rings, and the point submitted was whether on one or any of the hypotheses (1) that the rings are solid, (2) that they are fluid or in part uniform, (3) that they consist of masses of matter not mutually coherent, the conditions of stability are satisfied by the mutual attractions and motions of the planet and the rings.

Maxwell came to the conclusion that the third hypothesis was the only one compatible with the stability of the rings. It was a very heavy and difficult investigation and took two years' hard work. It added greatly to his reputation and showed that a new star of the first magnitude had risen in the firmament of Mathematical Physics. Airy said of it that it was one of the most remarkable applications of mathematics he had ever seen.

It was while at Aberdeen that he first turned his

attention to the Kinetic Theory of Gases. W. D. Niven, in his excellent memoir of Maxwell prefixed to the *Collected Works*, suggested that Maxwell had been led to do this by the problems he had met with in connection with his theory of Saturn's rings. Be this as it may, he gave in a paper read at the meeting of the British Association in Aberdeen the solution of the fundamentally important problem of the distribution of velocities among the molecules of a gas. This solution is known as the Maxwellian distribution and though his proof of it has been criticized no one doubts the truth of his result—like some other men of great physical insight, his instincts were better than his reasons.

When in 1860 the two Colleges, King's and Marischal, each of which had had a Professor of Physics, were fused into a University with but one Professor, the Professorship at Marischal College was suppressed. In the summer of 1860 he was appointed to the vacant Professorship of Natural Philosophy in King's College, London.

In February 1858 he had married Katherine Mary Dewar, the daughter of the Principal of Marischal College, and thereby ceased to be a Fellow of Trinity.

Before his election to the London Chair he had

been a candidate for the Professorship of Natural Philosophy at Edinburgh. Among the candidates were P. G. Tait and Routh, his old rival in the Tripos. Although Tait was elected, clear evidence of the position Maxwell had attained by this time is afforded by an article in the Edinburgh *Courant*, on the election, which says that "Professor Maxwell is already acknowledged to be one of the most remarkable men known to the scientific world", but it goes on: "there is another quality which is desirable in a Professor in a University like ours and that is the power of oral exposition proceeding on the supposition of imperfect knowledge or even total ignorance on the part of pupils. We have little doubt that it was a deficiency in this power in Professor Maxwell which made the Curators prefer the claims of Mr Tait". The view that Maxwell was unsuccessful as a teacher must seem strange to those who know of him only through his writings. His public lectures, such as the one on "The Dynamical Evidence for the Molecular Constitution of Bodies" before the British Association, are models of clear and brilliant expression, as are also his articles in the *Encyclopædia Britannica*. The public lectures were however read and he was compelled by the manuscript to keep to the track.

In class lectures to ignorant students, where the discourse would be much more informal, it is quite possible that he found it difficult to keep to the severely pedestrian tramp along one road necessary for such teaching and may have soared off into higher regions beyond their ken.

Maxwell began his work at King's College, London, in the autumn of 1860 and the five years during which he held the Professorship were perhaps the most fertile in his career. He produced in that time the paper on "The Theory of Colour", the two papers on "Physical Lines of Force" which with his paper on "The Dynamics of the Electromagnetic Field" contain his theory of Electricity and Magnetism, his Bakerian lecture on "The Viscosity of Air at different temperatures and pressures", as well as two important papers on "The Kinetic Theory of Gases". All his more important experimental work, with the exception of that with the colour box, was done during this period. The experiments on the viscosity of gases were made in a long attic at the top of his house in London, Mrs Maxwell acting as stoker and regulating the temperature. He also took an active part in the experiments to determine the Ohm in absolute measures which were carried out in the

Laboratory at King's College by a Committee consisting of Maxwell, Balfour Stewart and Fleeming Jenkin. He resigned the Professorship in 1865.

The years between his resignation and his return to Cambridge were for the most part spent at Glenlair. The greater part of the treatise on Electricity and Magnetism was written at this period and he also wrote a short treatise on Heat from which students of my generation learnt most of their Thermodynamics. He was President of Section A of the British Association in 1870, when his Presidential Address dealt with the relation between Mathematics and Physics. He examined in the Mathematical Tripos at Cambridge in 1866, 1867, 1869 and 1870. By the type of questions he set, and by the influence of his presence in Cambridge during the examinations, he was a principal factor in the movement which led to the introduction of the subjects of Electricity and Heat into the Mathematical Tripos and ultimately to the erection of the Cavendish Laboratory and to the foundation of the Chair of Experimental Physics.

Though Newton in his rooms in Trinity College and Stokes in his rooms in Pembroke had made physical experiments of fundamental importance, it was not until 1874 that the University possessed a laboratory in which physical experiments could be made or instruction in practical physics given.

In 1869 a Syndicate appointed by the University reported in favour of the foundation of a Professorship and Demonstratorship in Heat, Electricity and Magnetism and the provision of a Physical Laboratory, which they estimated would cost £6300. The University at first did not see how it could find the money for this, but at the beginning of the Michaelmas Term of 1870 the Chancellor of the University, the seventh Duke of Devonshire, who had been 1st Smith's Prizeman and Second Wrangler, offered to provide the funds for the building and apparatus as soon as the University had in other respects completed its arrangements for teaching Experimental Physics.

This eased the situation so much that in February 1871 the University approved the provision of a Professorship in Experimental Physics to which Maxwell was elected in March.

It is well to recognize on this occasion the debt which the University owes to the Duke of Devonshire; if it had not been for his generosity it would not have had the proud distinction of being able to claim Maxwell as one of its Professors, and though no doubt in time the University would have had a Physical Laboratory, it would have been without the inspiration and tradition which the Cavendish Laboratory owes to its first Director.

Maxwell's first work after his election was the preparation of plans for the new Laboratory. He threw himself into this with great energy and enthusiasm and spent much time and thought in planning the arrangements for teaching and research. The tender for carrying out these plans was considerably higher than the original estimate but the Duke undertook to defray the cost, and building began in the summer of 1872 and was completed by Easter 1874. In June of that year the Chancellor formally presented his gift to the University and expressed his wish to furnish the Laboratory completely with the necessary apparatus. In 1877 Maxwell reported to the University that the Laboratory contained all the instruments required in the present state of Science; this proved, as might be expected, to be unduly optimistic and Maxwell himself spent large sums in increasing the stock of apparatus; he was never content with any but the very best. In his inaugural lecture he set forth the part he thought the Laboratory ought to play in the studies of the University. I quote a few short extracts:

2-2

Such indeed is the respect paid to science that the most absurd opinions may become current provided they are expressed in language the sound of which recalls some well-known scientific phrase. If society is thus prepared to receive all kinds of scientific doctrines it is our part to provide for the diffusion and cultivation not only of true scientific principles but of a spirit of sound criticism.

The student who uses home-made apparatus which is always going wrong often learns more than one who has the use of carefully adjusted instruments to which he is apt to trust and which he dare not take to pieces.

He lectured each term during the building of the Laboratory, depositing his notions, as he said, like a cuckoo, one term in the Chemical Laboratory, the next in the Botanical, and then in the Museum of Comparative Anatomy.

The Laboratory was opened in 1874 with Mr Garnett as Demonstrator; the students were at first few in number, and for the most part men who had recently taken the Mathematical Tripos and who were interested in Physics. There were no organized demonstrations; the students were first set to make measurements with a few instruments, and after a short course of this kind began to work at a specific problem. W. M. Hicks was the first student, he was followed by J. E. H. Gordon, George Chrystal, Saunders, Donald MacAlister, Fleming, Glazebrook, Schuster, W. D. Niven, who edited Maxwell's *Collected Papers* and was the first to lecture in Cambridge on Maxwell's *Treatise*, and J. H. Poynting, who later made most important additions to Maxwell's theory, of which the Poynting Vector is now an integral part. Fortunately several of these are present at this Commemoration and we hope they will tell us of their experiences. One thing perhaps they may not tell us—that seven of these students became Fellows of Colleges and six Fellows of the Royal Society. What developed subsequently into a very important part of the work of the Laboratory, began at this period with the comparison by Chrystal and Saunders of the various units of resistance in the possession of the British Association. This was continued by Fleming and Glazebrook and was the beginning of researches which by the end of Lord Rayleigh's time covered the determination in absolute measure of all the fundamental electrical units.

Maxwell took great pains with the planning of a student's research; when this was done he acted on what I believe is the right principle, that it is best for the student to try to overcome his difficulties by his own efforts and that the part of the teacher is to encourage him to go on grappling with them rather than to remove them out of his way. He liked the students to have thought of a subject for themselves. "I never", he told Sir Arthur Schuster, "try to dissuade a man from trying an experiment; if he does not find out what he is looking for he may find something else."

The greater part of Maxwell's own work during this time was editing the papers of Henry Cavendish, who, though he published only two papers, had left twenty packages of manuscripts on Mathematical and Experimental Electricity. These Maxwell copied out with his own hand; he saturated his mind with the scientific literature of Cavendish's period, he repeated his experiments, being especially attracted by his device to use himself as a galvanometer to measure (Cavendish measured everything he came near) currents of electricity by the physiological effects they produced when they passed through him. Visitors to the Laboratory at this period had to submit to have currents sent through them to see whether they were good or bad galvanometers. *The Electrical Researches of the Honourable Henry Cavendish*, which was published in 1879 is unequalled as a chapter in the history of Electricity. It proved that Cavendish had anticipated several of the discoveries made after his time, that he had formed the conception of Specific Inductive Capacity as well as of Electric Capacity and had anticipated Ohm's law.

The small but very interesting book, *Matter and Motion*, containing his views on the principles of dynamics, was published at this period.

Each term he gave a course of lectures which did not attract as large an audience as they deserved.

The last few years of Maxwell's life were saddened and his attention occupied by the long and serious illness of his wife: he devoted as much of his time as was possible to nursing her, as he had done to his father in his last illness. At one stage of his wife's illness he did not sleep in a bed for three weeks but gave his lectures and did his work at the Laboratory as usual.

At the beginning of 1879 his friends noticed a serious turn for the worse in his health and in the Easter Term, though he attended the Laboratory daily, it was only for a very short time. In June, after his return to Glenlair, he rapidly got worse and in October he was told that he had not a month to live. He returned to Cambridge to be under the care of his favourite physician, Dr Paget. Though he suffered great pain which he bore with wonderful fortitude, his mind was absolutely clear, his one thought was for the welfare and comfort of his wife who had been accustomed to rely upon his help in almost every detail of her life. She was now bedridden and to the last Maxwell gave the orders necessary for her comfort. The end came on November 5 when he was in his forty-ninth year. After a memorial service in the Chapel of Trinity College, of which he was an Honorary Fellow, the body was taken to Glenlair and buried in Parton churchyard.

In preparing this part of my address I have been greatly indebted to Lewis Campbell's *Life of Maxwell* and the more I studied that book the more has the impression which I long ago got from conversation with those who had the privilege, which I had not, of knowing Maxwell been strengthened and deepened. That impression is one of profound admiration for the fineness and strength of his character, for his unselfishness and kindness. He is revealed as one of the most lovable of men, a good son, a good husband, a good friend, a lover of literature, of philosophy and of theology.

He was a deeply religious man, very reticent in speaking about religion to any but those closest to him; the remarkable letters to his wife printed in the *Life* show how deep this feeling was, and his views about the expression of that feeling by the letter which he sent in reply to an invitation to join a Society for the discussion of religious questions by men of science. He says: "I think the results which each man arrives at in his attempts to harmonize his science with his Christianity ought not to be regarded as having any significance except to the man himself and to him only for a time".

Though he held the Professorship for but eight years, he did a great work in that time; he planned the building, stocked it with apparatus and made it available for the study of Physics. Most important of all, he settled the policy of the Laboratory, he breathed into it the spirit of research, he determined that it should create new knowledge as well as teach the old. He set by his own work and that of his students a high ideal for those who work in the Laboratory and left us the proud heritage of a great name.

How great that heritage is was not realized until long after his death. The truth of his supreme contribution to Physics—the theory of the Electric Field—was at the time of his death an open question. There was not much experimental evidence in its favour, though there was none against it. The most that could be claimed for it was the Scottish verdict non-proven. It got very little support outside a small group of young Cambridge men. The reason it was not adopted by the older men was not hide-bound conservatism or wilful blindness to evidence. They had been working for years with the old theory, they had been led by it to great discoveries, they knew it was not inconsistent with any known electrical phenomena. For them to forsake this theory for one which introduced a new principle for which there was no direct evidence was quite a different thing from the adoption of a new theory by young men who were just beginning to study electricity and who had no theories to give up. I hope I shall be pardoned if I devote a short time to this part of Maxwell's work.

He began the study of Electricity in 1854 when the science was in a very anomalous condition. On the one hand there were the mathematicians, who viewed the subject entirely from the point of view of action at a distance. They supposed that the space outside an electric charge or a magnetic pole was just the same as if these were not present, that outside them there was nothing but distance. From this point of view they solved many problems, such as those relating to the distribution of electricity on bodies of various shapes which were interesting from the elegance and ingenuity of the methods employed but not very fundamental from the physical point of view. On the other hand there was Faraday, who under the inspiration and guidance of an entirely different outlook had for more than twenty years been making time after time discoveries which opened up entirely new regions of knowledge. To Faraday the charge or pole was the starting-point of lines of force spreading out in every direction; thus the space around them was not as other space since it was filled with these lines. On Faraday's view it was these lines of force which were responsible for electric and magnetic actions—they were the things which really mattered. To him they were not merely geometrical lines, they had physical properties, they were in a state of tension and this produced the attraction between charges of opposite signs; again, the energy in the field gathered round them. Though Faraday showed by the way he handled his conception of the lines of force that he was a born geometer he was nothing of an analyst and never attempted to put his views into a mathematical form; there is not an algebraical

symbol in the *Experimental Researches*, nor indeed is there in Newton's *Principia*. The mathematicians could not understand him and were not
a little contemptuous. "I declare", said the Astronomer Royal of the time, "I can hardly imagine
any one who knows the agreement between observation and calculation based on action at a distance
to hesitate an instant between this simple and
precise action on the one hand and anything so
vague and varying as lines of force on the other."
The agreement referred to relates to problems like
those to which we have just alluded. Maxwell in
his paper on "Faraday's Lines of Force" 1855, the
first he wrote on Electricity, showed that for such
problems Faraday's views led to exactly the same
results as the theory of action at a distance.

But while these questions were the only ones
within the purview of that theory, they were only
a part and not the most important part of those
within the range of Faraday's method. This was
constantly suggesting the possibility of new phenomena, and continually raising new questions. For
example, Has the space traversed by the lines of
force the same properties as empty space? Faraday's great discovery of Electromagnetic induction
showed that it had not, for if in space through

which lines of magnetic force are passing a wire circuit is twisted electric currents will flow through it, but will not do so in normal space. Faraday arrived at the conception that the space traversed by lines of magnetic force was in a special state, which he called the *electro-tonic* state, in which it exhibits properties foreign to normal space. The conception of lines of force is in my opinion one of the greatest of Faraday's many great services to Science and I think that the properties of the electric and electromagnetic fields find their simplest and most suggestive expression in terms of them.

There is a very striking analogy between lines of force in the electromagnetic field and vortex filaments in a liquid whose boundaries are fixed. The motion of the liquid and its energy depend entirely on the positions and strengths of these filaments; they are so to speak the co-ordinates by which the motion of the fluid can be determined, just as the lines of force are the co-ordinates which determine the behaviour of the electromagnetic field. Again the vortex filaments cannot end in the liquid itself, they must either go to a boundary or else form closed curves. The lines of force also cannot end in free space, their ends must be on electric charges or they must form closed curves. And yet again the vortex filament, like lines of force, can neither be created nor destroyed; if at any place there is a change in the number of filaments it must be because filaments have drifted into or away from that place. The equations which determine the motion of a fluid in which there are a great number of vortex filaments moving, like the molecules of a gas, in all directions, are of the same form as Maxwell's equations of the electromagnetic field, so that such a fluid can be used as a model of the field.

The next paper published by Maxwell on Electrical Theory was the one on "Physical Lines of Force". This paper is of very great interest, for we can trace in it the development of the ideas which are his greatest contribution to Electrical Theory. His first paper had been more a translation of Faraday's views into mathematical language than the introduction of a new principle. In the paper on "Physical Lines of Force" we see the new principle gradually being built up with all the scaffolding around it. In his *Treatise* the scaffolding is pulled down; it had done its work but as he himself said "Science is most easily digested when in the nascent state", and I think some will find it easier to get at his meaning in the paper than in the treatise. In the paper his method is to devise a model of the magnetic field which will illustrate Faraday's law of electromagnetic induction. No one ever appreciated more than Maxwell the advantages gained in concentration of thought and in the suggestions of new ideas by considering a concrete case, like that of a model, instead of relying upon symbols. He says: "For the sake of persons of different types of mind scientific truth should be presented in different forms and should be regarded as equally scientific whether it appears in the robust form and vivid colouring of a physical illustration or in the tenuity and paleness of a symbolical expression". Boltzmann says: "Perfect elegance of expression belongs to the French, the greatest dramatic vigour to the English, above all to Maxwell".

In Maxwell's model the lines of magnetic force were represented by cylinders rotating round these lines as axes, the magnitude of the force being represented by the velocity of rotation and its direction by that of the axis of rotation.

In a uniform magnetic field the cylinders would all have to be rotating in the same direction. The question was how they were to be geared together to do this. If two adjacent wheels were in contact they would rotate in opposite directions; to make

them rotate in the same direction he introduced
between them small spheres like ball-bearings to
act as idle wheels; these would rotate in opposite
directions to each of the cylinders with which they
were in contact and so these would rotate in the
same direction. Maxwell supposed that these balls
represented particles of electricity and their motion
an electric current. When the wheels are rotating
with the same velocity the ball-bearing between
them will remain in the same place, merely rotating about their axes; since they have no motion
of translation there is no current. But suppose
the speed of one of the cylinders is changed; this
corresponds to a change in the magnetic force;
this cylinder will no longer rotate with the same
speed as its next neighbour and the centre of the
ball between them will move. Since the motion
of the ball indicates an electric current, the model
illustrates the production of electric currents by
changes in the magnetic form, i.e. Faraday's discovery.

Suppose now that the cylinders are at rest and
a force is applied to the balls; as the balls move,
the cylinders in contact with them will rotate, and
the rotation of the two in contact with a ball will
be in opposite directions. The rotation of the cy-

linders indicates a magnetic force. Thus the model suggests that if you move the electricity you get magnetic force. In a conductor of electricity the electricity when acted on by an electric force can continue in motion and thus the wheels are kept rotating and the magnetic force persists. In an insulator the electric charge will begin to move when the force is applied but the motion will be resisted and finally stop; as long however as it is moving there will be rotation of the cylinders and magnetic force. Any alteration in the electric force will alter the positions of equilibrium of the balls, and as the balls move from one place to another the cylinders will rotate and magnetic force be produced.

This is a very striking illustration of the advantages of using a model. Maxwell designed a model to illustrate Faraday's discovery that changes in the magnetic force produce electric forces; when he came to use the model he found that it suggested that *changes* in the electric force would produce magnetic force. The introduction and development of this idea was Maxwell's greatest contribution to Physics. The importance of the step made by Maxwell is indicated by the fact that on the electromagnetic theory which held the field before his time, electrical waves could not exist, while on his theory all changes in electric and magnetic forces sent waves spreading through space.

He returns to the theory in the latter part of a later paper "The Dynamics of the Electromagnetic Field". The model used in the paper on "Physical Lines of Force" is not introduced; it had done its work by suggesting the existence of these new currents and the modifications they involved in the equations connecting electric and magnetic force. In the later paper he postulates the existence of these currents and develops their consequences; he obtains equations which are in substance the well-known Maxwellian Equations. These as a matter of fact had been obtained though in a more sophisticated form in the paper on the "Physical Lines of Force", which is a most fascinating account of the birth of a theory. I remember that when I was a boy of eighteen I was raised to such a pitch of enthusiasm by reading this paper that I copied the whole of it out in longhand and it is a very long paper.

Maxwell's final presentation of his theory is given in the *Treatise on Electricity and Magnetism* published in 1873. The greater part of this was written at Glenlair in the interval between his leaving King's College and his return to Cambridge. It is a general treatise on Electricity and Magnetism and by no means confined to the discussion of the theory. At first the theory had few adherents. One reason for this, it must be confessed, is that his presentation of his theory is in some features exceedingly obscure. One of the most difficult chapters in the literature of Physics is that at the beginning of the first volume in which he describes the features peculiar to the theory. Most important of these is the requirement that wherever there is electric force there is what he calls a "displacement" of electricity; he draws an analogy between this and the displacement of an elastic solid under stress. This led many to suppose that he regarded electricity as a kind of elastic solid and displacements of electricity as the displacements of this solid. He says that the movements of electricity are like those of an incompressible fluid so that the total quantity within a closed surface always remains the same; this shows that what he meant by electricity was something different from a collection of electric charges. In the paper on "Physical Lines of Force", in which he built up his theory, he regards electricity as consisting of small par-

3-2

ticles, and the displacement of electricity the motion of these in a molecule, yet he never alludes to this in the *Treatise* and when he is driven for convenience of description in discussing electrolytic phenomena to speak of a molecule of electricity he says the phrase is out of harmony with the rest of the *Treatise*. In fact he is so guarded about assuming anything about electricity that he postulates little about it beyond its name. This is not enough to enable us to visualize the part it plays in electrical phenomena. This difficulty in visualizing what Maxwell meant was one of the reasons why the theory was so slow in gaining acceptance; it was this that was in Hertz's mind when he said that Maxwell's theory was Maxwell's equations and in Helmholtz's when he said that he would be puzzled to explain what an electric charge was on Maxwell's theory beyond being the recipient of a symbol.

Fortunately we have the material for such a physical interpretation of Maxwell's theory if we go right back to Faraday and his lines of electric force. If we regard what Maxwell called electricity as being really lines of force, and what he called the displacement of electricity as the density of these lines, this view satisfies the two definite statements about electricity made by Maxwell: (1) that wherever there is electric force there is electric displacement; (2) that electricity behaves like an incompressible fluid. Maxwell in his first paper had shown that the lines of force coincided with the lines of flow of such a fluid.

Again, lines of force are not created afresh, so that when the number passing through unit area (in Maxwell's notation the "electric displacement") varies, there must be a motion of the lines of force to or from this area, and the rate of change in the number passing through a closed circuit must equal the number passing across its boundary in unit time. When the lines of electric force move they produce magnetic force as is shown by the fact that a moving electric charge is surrounded by a magnetic field. If the magnetic force due to the motion of the lines of electric force is at right angles to these lines and also to the direction in which they are moving, and proportional to their velocity at right angles to their length, the work done in taking a unit magnetic pole round the boundary of a circuit is proportional to the number of lines of force crossing the boundary in unit time and therefore to the rate of change in the number of lines of electric force passing through the circuit. This result is the essential feature of Maxwell's theory. The lines of electric force were introduced by Faraday to represent the processes going on in electrostatic phenomena. In these the electric force does not vary with the time and the lines of force are at rest. In by far the greater number of electrical phenomena the lines of electric force are in motion and not at rest. The motion of the lines of force gives them new properties and it might well have been that these might have failed entirely to represent the phenomena of electrodynamics, though their properties when at rest sufficed for electrostatics. It is strong evidence of the fundamental character of the conception of lines of force that the properties they acquire when moving are exactly those which are required to enable them to represent the phenomena of electrodynamics.

Faraday's law of electromagnetic induction is that the rate of alteration in the number of lines of magnetic induction passing through a circuit is equal to the work done in taking unit electric charge round the circuit.

Maxwell's law is that the rate of alteration in the number of lines of electric force passing through a circuit is equal to the work done in taking a unit magnetic pole round it. The expression of these two laws in a mathematical form gives the system of equations known as Maxwell's equations. You will notice the symmetry of the laws: if we interchange the words electric and magnetic in Faraday's law we get Maxwell's.

The most striking consequence of Maxwell's theory is that electric disturbances are propagated as transverse waves of electric and magnetic force. The velocity of propagation of these depends upon quantities which can be determined by purely electrical measurements, the velocity so calculated being, within the limits of experimental error, equal to that of light. This of course suggested that waves of light are waves of electric and magnetic force, which is Maxwell's Electromagnetic Theory of Light. Light waves however are, if Maxwell's theory be true, only a small fraction of the electrical waves, which though we cannot see them must always exist in the space around us. The detection of these is vital for the establishment of the theory. Maxwell himself did not make any experiments to test his theory. During his life he had not had many opportunities of making elaborate experiments. When at Aberdeen his time was occupied in working for the Adams Prize, a long and difficult piece of work which had to be finished by a definite date. When at King's College he took advantage of his opportunities and did a considerable amount of experimental work; the electromagnetic theory however had only just been developed in 1865 when he left the College. From 1865 to 1871 he was at Glenlair and had no opportunities for experiments, and though he returned to Cambridge in 1871 the Laboratory was not opened until 1874 and then he was fully occupied with editing the Cavendish Papers, a pious duty for a Cavendish Professor. His strength too was in Theoretical rather than Experimental Physics.

It was not until a long time after the publication of the *Treatise* that any method for producing and studying electrical waves in the laboratory was discovered. We had to content ourselves with the ready-made ones occurring in light. The only part of Maxwell's theory which could be tested was thus his Theory of Light. It bore the test well. I have already said that it gave the right velocity for light waves in empty space. It indicated that the velocity of light waves through matter would be inversely proportional to the square root of its specific inductive capacity. In some cases this was found to be true, in others it was very wide of the mark. The divergence was not surprising; what was surprising was that there should be any substance for which it was true, for the specific inductive capacity had been measured under an unchanging electric force, while the force in the light waves changes backwards and forwards millions of millions of times per second. A better test was to find how on this theory the intensity of reflected and refracted polarized light varied with the angle of incidence and the plane of polarization, what would be the laws of scattering of light by small particles, and so on. The results of such tests were distinctly in favour of the theory, the agreement between theory and experiment being on the whole better than for any other theory of light.

It was not however until nearly ten years after Maxwell's death that any direct experimental evidence of the existence of electrical waves was obtained. For this two things are necessary: (1) a method of producing these waves if they exist, (2) a means of detecting them when produced. The oscillatory discharge of a condenser furnished the first; the difficulty was to detect them. This difficulty arose because the electric currents in the waves were changing their directions millions of

times per second. There were then no methods
available for detecting currents alternating as
rapidly as this; now it is easy to get a circuit
which offers only a one-way traffic to the current, such as a circuit containing a wire in contact
with certain crystals or a thermionic valve. These
methods had not been discovered at that time and
the actual discovery of electric waves was made
by means of the oldest, in order of discovery, of
electrical phenomena—the electric spark. The
spark which passes between two metal balls placed
close together, though it requires a considerable
force to produce, is, when such a force is available, produced in a very short time, under fa-
vourable conditions much less than a millionth of
a second. The electric force in the waves is acting
in one direction for longer than this and thus
can produce its effect before a force in the opposite direction comes to interfere with it. It was
by using these sparks and observing the alter-
nation of the length of the spark when the spark-
gap was placed in different positions that Hertz,
a pupil of Helmholtz (who was the first Continental
physicist to support Maxwell's theory), established
the existence of electrical waves. He showed that
they were reflected in the same way as light waves,

hat they could be brought to a focus, that like
ght they could be polarized and most important
f all for establishing their wave character showed
nterference effects from which the lengths of the
vaves could be calculated.

Hertz's researches were one of the most mar-
rellous triumphs of experimental skill, of in-
;enuity, of caution in drawing conclusions, in the
vhole history of Physics. Younger physicists, with
:he very efficient means of detecting electrical
vaves now at their command will not realize the
lifficulties of these experiments, but older ones, who
ike myself began by using Hertz's method and
had to observe whether tiny sparks only a fraction
of a millimetre long waxed or waned when the
detector was moved from one position to another,
will remember how arduous and harassing these
experiments were and how long it took to make
sure that the effects observed were not spurious.
Rough as the method seemed, it was able in Hertz's
hands to prove the existence of electrical waves
and to enable him to establish the theory of which
Maxwell was the author; the two names Maxwell
and Hertz will always be associated in the history
of this subject.

The discovery of electrical waves has not merely

scientific interest though it was this alone which inspired it. Like Faraday's discovery of electromagnetic induction, it has had a profound influence on civilization; it has been instrumental in providing methods which may bring all the inhabitants of the world within hearing distance of each other and has potentialities social, educational and political which we are only beginning to realize.

MAXWELL'S INFLUENCE ON THEORETICAL PHYSICS IN GERMANY

BY

Max Planck

F.M.R.S.

A GREAT investigator influences the intellectual world in the first instance through his scientific results. These are the most immediate and the most valuable fruits of his life's work. Often, however, a great personality will wield an influence of a less direct kind, but one often of comparable importance. This may arise through the stimulating influence exerted by such a man on a group of sympathetic contemporaries, through whose agency his scientific gifts may bear further fruit. In the field of the sciences of the mind it is not always that this distinction between direct and indirect influence can be sharply made, since a moulding of cultured opinion may well be a main part of his endeavour. But in science, where the worker and the object of his investigation are clearly distinguished, it is much easier to follow in detail the way in which a great investigator may achieve immortality, not only through his own discoveries but also through those that he inspires others to make.

It must be admitted, however, by anyone who believes that the Physical Sciences are concerned with the description of an objective world rather than of a private world of personal experiences, that the contact of learned men of diverse countries is practically unnecessary. Even though all the countries of the earth were isolated from each other, the development of physics would everywhere take the same course. This view is supported by the fact that, once the necessary objective conditions are fulfilled, great physical and technical discoveries are often made independently in different countries. Up to a point therefore scientists of different lands carry out their work independently of each other.

There are however in every science certain exceptional individuals, who appear divinely blest, and radiate an influence far beyond the borders of their land and thus directly inspire and expedite the research of the whole world. Among these is to be counted *James Clerk Maxwell*, the centenary of whose birth is now being celebrated. Although one cannot doubt that all that he achieved over the whole field of physics would sooner or later have become the common knowledge of science, even had he never lived, yet to him belongs the honour not only of so many first discoveries, but of having served so well his contemporaries of all lands by his stimulating influence and by having saved them from many arduous and useless bypaths.

That Maxwell's great discoveries were in no sense accidental, but that they arose out of the abundance of his genius, is shown by the many fields in which he was pioneer, leader and master.

Modern physics recognizes two main conceptual schemes, the physics of discrete particles and the physics of continuous media, and it is since Maxwell's time that the distinction between them first became more apparent. These schemes correspond nearly but not quite to the physics of Matter and the physics of the Aether. In both regions Maxwell introduced new and fruitful ideas.

An attempt to show the significance of these conceptions for the development of physics in Germany will be most clearly made by studying the influence exerted by Maxwell on those of his German fellow scientists, who ranked at that time or soon after as leaders of their science.

The physics of Particles, to consider this first, had its origin in the distant past, but was reborn about the middle of the last century in the form of the kinetic theory of gases. This theory, which followed closely on the discovery of the mechanical equivalent of heat, was formulated simultaneously by independent workers in more than one country; by J. P. Joule and J. J. Waterston in England and by A. Krönig and R. Clausius in Germany. Maxwell, too, interested himself early in this new hypothesis, which attributed both pressure and heat of a gas to the rapid irregular motion of swarms of flying molecules, colliding continuously with each other and with the walls of the container. So bold and surprising did the hypothesis then appear that it was violently attacked by every sort of positivist as a dangerous error. To the simple relations connecting the mean velocity of the molecules with the pressure and specific heat, which had been derived by his predecessors, Maxwell added a new and more fundamental set of considerations, by enquiring as to the actual velocity of any molecule selected at random. By his answer to this question Maxwell laid the foundations of a new branch of physics, that of Statistical Mechanics. For it is clear that the question can only be answered by a probability law, that is by a law which gives the fraction of the molecules which are found to possess a definite velocity when the experiment of selecting a molecule at random is repeated a great number of times. This law, first found by Maxwell and named after him, was seen to be identical with the Gaussian Error Law, so long at any rate as the three components of the velocity vectors could be considered as independent.

The effect of this discovery in Germany was very varied. Krönig does not appear to have concerned himself with this question. Clausius, while admiring the result, clearly attributed no great importance to it, since he considered and attempted to prove that its validity was limited to the case of elastic molecules treated by Maxwell.

But with Ludwig Boltzmann it was quite otherwise. He recognized at once and with great clearness the fundamental position of Maxwell's Velocity Distribution Law in the Kinetic Theory of Gases. In so doing he prepared the way for the acceptance of Maxwell's work in Germany, although, or perhaps because, he was a keen critic of it.

Boltzmann first refined and generalized Maxwell's proof, which assumed monatomic spherical molecules to cover the case of polyatomic molecules. He showed further by means of his well-known H-theorem that not only does the Maxwell distribution remain stationary, once it is attained, but that it is the only possible equilibrium state, since any system will eventually attain it, whatever its initial state.

Boltzmann showed further that with every degree of freedom of a molecule of a gas in a steady state is associated the same average energy. A difficulty which had been encountered by Maxwell when calculating the specific heat, and which threatened to prove embarrassing to the kinetic theory, was also successfully overcome by Boltzmann. This concerned the ratio of the specific heat at constant pressure to that at constant volume, a ratio which plays an important part in all adiabatic processes. While for a monatomic gas, such as mercury vapour, this ratio is observed to have exactly the value of $1\frac{2}{3}$ appropriate to the assumption of spherical molecules, for polyatomic molecules a definite discrepancy was found between theory and experiment. For if three different moments of inertia are attributed to a non-spherical molecule, the ratio of the two specific heats should be $1\frac{1}{3}$, while the observed values for hydrogen, oxygen, and nitrogen was $1\frac{2}{5}$. Boltzmann discovered a simple way out of the difficulty by introducing the assumption that such molecules have only two instead of three different moments of inertia. This assumption is in agreement with the fact that these molecules are diatomic, so that the line joining the two atoms is a symmetrical axis of rotation.

The further question as to how that degree of freedom behaves, which corresponds to the mutual vibration of the atoms of a molecule, could be answered neither by Boltzmann nor Maxwell; for the solution of this question belonged to a later epoch of physics.

Thus we see how these two investigators, each stimulating the other by friendly rivalry, together built up the young science of statistical mechanics. It is peculiarly attractive to follow them as they both strode forward along the different roads mapped out for them by their individual temperaments, and yet to watch them checking and extending each other's work, until they finally met at the same goal. The difference of method can be seen, for instance, in the fact that Maxwell attempted to find the statistical laws of a complicated mechanical structure by considering the structure simultaneously in a great number of different states, while Boltzmann on the other hand preferred to follow over a long period of time a single structure through its manifold changes of state. Both methods, if carried through consistently, lead to the same statistical laws. Both men were quite clear as to the close connection between statistical mechanics and thermodynamics, and both were agreed that the second law of thermodynamics is a probability law, and must therefore, from the point of view of mechanics, allow particular exceptions.

The kinetic theory encountered great difficulties in connection with the way in which such irreversible processes as viscous flow, diffusion and conduction of heat varied with the time. Even though certain conclusions of the theory, such as Maxwell's deduction that the viscosity of a gas is independent of its pressure, were in excellent agreement with observation, the attempt to calculate exactly the numerical value of the coefficient of viscosity proved very difficult. For, in order to be able to carry out the very complicated calculations, certain simplifying assumptions were necessary, as for instance that the velocities of all the molecules are the same, or the more essential assumption that the distribution of velocities in a streaming gas is obtained by superimposing the stream velo-

city on to the velocity distribution for a gas at
rest. But all such assumptions led to contra-
dictions, since among the quantities neglected were
always some of the same order as those taken into
account. Each of some six or more investigators in
this field obtained different values for the ratio of
the coefficient of diffusion to the coefficient of conduction of heat, according to the particular method
of calculation.

Boltzmann discovered, in principle, the way to
escape from this labyrinth, since he succeeded in
obtaining an exact expression for the velocity distribution in a layer of gas which is not in a steady
state. The difficulty, however, arose that it proved
impossible to satisfy this equation, at any rate for
the simplest case of elastic and spherical molecules.
With the perseverance and determination which
characterized him, Boltzmann dedicated a considerable, perhaps even too considerable a part of
his invaluable powers to the attempt to solve this
problem by means of expansion in series, first in
one way, then in another. The pages of his three
papers, "Zur Theorie der Gasreibung", are covered
with almost endless formulae and figures and give
impressive evidence of the long and troublesome
calculations.

Maxwell proceeded otherwise. Instead of allowing himself to get held up, as did Boltzmann, by the intractable problem of finding a solution for the case of elastic spherical molecules, he altered the problem by substituting for the elastic molecule others with more convenient properties. He was led to the possibility of this trick by the view that the properties of the pressure and the viscosity, etc., of a gas must be widely independent of the particular laws which govern the collision of two molecules, so long as the conservation of energy and momentum holds, since a collision takes place in a relatively very short time. In the case of hard elastic bodies, the collision is a completely discontinuous process, since the molecules retain their velocity unchanged in magnitude or direction till at the instant of collision they are suddenly altered. Instead of this, and so long as one is concerned only with the final result, one can consider a collision as a continuous but rapid transition from the initial to the final velocity, by assuming a repulsive force between the molecules, whose magnitude is inversely proportional to a not too low power of their mutual distance. Then the molecules act almost independently when far away from each other, that is, they move with constant velocities, and only when close together do they suffer such large changes in velocity as to constitute a collision.

Among all the possible power laws of force, that of the inverse fifth power proved particularly convenient. For this law the closest distance of approach of two molecules during a head-on collision is inversely proportional to the fourth root of their relative velocity, and it follows that the relative velocity of the molecules disappears from the final expression for the viscosity, so that a general expression for the velocity distribution of a gas not in a steady state is not needed. Maxwell therefore postulated a repulsive force varying as the inverse fifth power of the distance, and so solved comparatively simply the problem of viscosity.

Both this achievement of Maxwell's and the form of its presentation made such an impression on Boltzmann that he ranked it as a perfect work of art. In sublime words he compared Maxwell's work to a great musical drama. This he sketched out in a way almost as equally characteristic of him as of Maxwell.

At first are developed majestically the Variations of the Velocities, then from one side enter the Equations of State, from the other the Equations of Motion in a Central Field; ever higher sweeps the chaos of Formulae; suddenly are heard the four words: "put $n = 5$". The evil spirit V (the relative velocity of two molecules) vanishes and the dominating figure in the bass is suddenly silent; that which had seemed insuperable being overcome as if by a magic stroke. There is no time to say why this or why that substitution was made; who cannot sense this should lay the book aside, for Maxwell is no writer of programme music, who is obliged to set the explanation over the score. Result after result is given by the pliant formulae till, as unexpected climax, comes the Heat Equilibrium of a heavy gas; the curtain then drops.

We will also let the curtain drop and turn to another region of physics, where Maxwell's genius achieved yet greater triumphs: the physics of Aether or Electrodynamics. While, in the kinetic theory of gases, Maxwell shared his leadership with several others, in the field of electrodynamics his genius stood alone. For to him was given, after many years of quiet investigation, a success which must be numbered among the greatest of all intellectual achievements. By pure reasoning he succeeded in wresting secrets from nature, some of which were only tested a full generation later, as a result of ingenious and laborious experiments. That such predictions are at all possible would be quite unintelligible if one did not assume that very close relations exist between the laws of nature and those of the mind.

We must not of course forget that Maxwell did not build his Electrodynamic Theory in the air. For out of nothing comes nothing. He built his theoretical speculations on the firm foundations laid by the experimental work of Michael Faraday, whose memory we have so recently and so fittingly celebrated. But Maxwell, with bold phantasy and mathematical insight, went far beyond Faraday, whose standpoint he both generalized and made more precise. He thus created a theory which not only could compete with the well established theories of electricity and magnetism but surpassed them entirely in success. For the criterion of the value of a theory, that it explains quite other phenomena besides those on which it was based, has never been so well satisfied as with Maxwell's theory. Neither Faraday nor Maxwell may have originally considered optics in connection with their consideration of the fundamental laws of electromagnetism. And yet the whole field of optics, which had defied attack from the side of mechanics for more than a hundred years, was at one stroke conquered by Maxwell's Electrodynamic Theory; so much so that since then every optical phenomenon can be directly treated as an electromagnetic problem. This must remain for all time one of the greatest triumphs of human intellectual endeavour.

However, Maxwell's theory had at first a somewhat difficult career, due mainly to its original character. What made it so difficult of general comprehension and what therefore so greatly lessened its force of conviction, was the impossibility of devising a visualizable model which would relate its formulae to ordinary mechanical principles.

In Germany this difficulty acted as a grave impediment. Here in the middle of the last century the completion of Electrodynamics was sought entirely in terms of Potential Theory, which had been derived by Gauss from Newton's law of action at a distance for statical electric and magnetic fields and which had been brought to a high degree of mathematical completion. The generalizations required to include dynamical processes were sought through an extension of Newton's law of gravitation, in that the force of attraction of two masses could be assumed to depend on their velocities and accelerations as well as on their positions.

The assertions of Faraday and Maxwell that such immediate action at a distance did not exist at all and that the field of force had an independent physical existence, were so foreign to this whole

method of thought that Maxwell's theory found no
foothold in Germany and was scarcely even noticed.
If noticed at all, it was considered as an interesting
curiosity. Only a few physicists felt disposed to
approach nearer. Among these was Ludwig Boltzmann, who studied particularly Maxwell's asserted
relation between the refractive index and the dielectric constant, and verified it completely by
extremely careful experiments on various sub-
stances, especially on gases. Less successful were
his repeated attempts to make Maxwell's electro-
dynamical equations more intelligible by means of
mechanical models.

H. von Helmholtz, who approved Maxwell's
theory on account of its remarkable formal sim-
plicity, took a middle course. Through important
investigations he succeeded in setting up a general
law for the interaction of open electric circuits,
which gave as special cases both the various action-
at-a-distance theories and the corresponding formula of Maxwell. But this method could not remove
the fundamental antithesis between action at a
distance and by immediate contact. The final de-
cision in Germany and in the whole world in this
conflict of theories was made in favour of Maxwell
by Heinrich Hertz, the most distinguished pupil of

Helmholtz. It is remarkable that many years earlier than his epoch-making experiments, Hertz had been led by theoretical considerations, based on the then known facts of physics, to the view that Maxwell's theory was superior to those of action at a distance. Since his argument does not appear to have received its due recognition, it will be described shortly here.

If there is only a single kind of electrical force, so that the force with which a rubbed ebonite rod attracts or repels a pith ball is the same force by which a moving or varying magnet induces an electric current in a conductor, then the same magnet should set a pith ball in motion; according therefore to the mechanical principle of action and reaction, an electrically charged body should thus produce a ponderomotive force on a moving magnet, and so finally, quite apart from the usual magnetic action, one moving magnet must act ponderomotively on another, with an electric force which depends on the relative motion of the magnets. But the electromagnetic theory which was based on action at a distance assumed only such ponderomotive forces between magnets which depend only on their instantaneous magnetization and not on their time variation; consequently from this point of view such an electric dynamics could not be complete.

The addition of the necessary term gives a correction which is certainly very small, since it contains the square of the so-called critical velocity in the denominator. One cannot however stop there. From the correction of the ponderomotive action, one unavoidably obtains, according to the principle of conservation of energy, a correction to the inductive action. Since however the induction and the ponderomotive forces are essentially the same, a new correction to the ponderomotive force is required, and so on for ever. If one introduces each time the appropriate correction, one clearly obtains, both for the ponderomotive effect and for the electric and magnetic inductive effects, an infinite series of terms which contain decreasing even powers of the critical velocity and which thus, in general, converges. The remarkable fact appears that this series accurately satisfied the differential equations set up by Maxwell according to which disturbances are propagated with the critical velocity.

This peculiar derivation of the Maxwell theory from the assumption of an immediate action at a distance was naturally not considered by Hertz as a proof of the correctness of the theory, because one cannot draw certain conclusions from uncertain assumptions, but it was considered as sufficient for the following conclusion. "If the only choice lay between the usual system and Maxwell's, then *the latter has undoubtedly the advantage.*"

By a curious coincidence this paper of Hertz appeared just as Maxwell's theory received a powerful support through the short but now famous contribution of Boltzmann on the temperature variation of the heat radiation of a black body. In this he derived the empirical law of J. Stefan from the second law of thermodynamics, using Maxwell's expression for the pressure of radiation.

Thus gradually the universal significance of Maxwell's ideas became to be more and more recognized on all sides, till at last the crucial experiments of Heinrich Hertz with very rapid electrical oscillations were crowned with an unexampled success, by the production of electrical waves of a few centimetres wave-length. Through this discovery, which produced the greatest sensation in all the scientific world, the speculations of Maxwell were translated into fact and a new epoch of experimental and theoretical physics was begun.

The meaning of Hertz's experiments for Maxwell's theory appears yet more striking, when one remembers that Hertz by no means only set out to verify Maxwell's theory. How little he was prejudiced from the theoretical side is clear from the fact that, for a time, he believed that he had established by his experiments a difference between the velocity of propagation of electrical waves in free space and along wires, a result which was in disagreement with Maxwell's theory. Only later did he recognize that the observed difference was due to the disturbing influence of surrounding conductors.

From now onwards the victory of Maxwell's theory was conclusive and the next problem of research concerned its further extension in all possible directions, particularly in the production and investigation of waves of such wave-length as to lie in the region between optical and electrical waves.

Among the German physicists who contributed to this work, it was above all Heinrich Rubens who proved, in conjunction with Ernst Hagen, that the measured reflexion of light by metals agreed in all details with Maxwell's theory, as soon as one employed light of sufficiently great wave-length. Since Maxwell himself had seen great difficulties for his theory in this phenomenon, this verification changed a great difficulty into a great success.

There remained outstanding as a dark spot the question of the reflexion of short wave-length light by metals. Here we reach, in fact, the limit, beyond which the Maxwell equations in their original form could not go, based as they were on the assumption of continuously distributed matter; the necessity of introducing atomic processes soon became evident. As time went on and as progress was made in the technique of measurements, it became clear that it was not matter alone which possessed atomic properties, but that energy too in a certain sense possessed them also; further it began to be seen that even the distinction between corpuscular and wave processes, a distinction which till then had been introduced into physics as a matter of course and which has also formed the basis of the disposition of these reflexions, cannot be accomplished completely, but can only be admitted as a limiting case. For just as on one side the energy in a homogeneous wave appears always in discrete particles, so on the other the collision of two molecules must be treated as the interference of two wave groups.

Maxwell did not live to see this revolutionary

development; it was his task to build and complete the classical theory, and in doing so he achieved greatness unequalled. His name stands magnificently over the portal of classical physics, and we can say this of him: by his birth James Clerk Maxwell belongs to Edinburgh, by his personality he belongs to Cambridge, by his work he belongs to the whole world.

MAXWELL'S INFLUENCE
ON THE DEVELOPMENT OF THE
CONCEPTION OF PHYSICAL REALITY

BY

Albert Einstein

THE belief in an external world independent of the percipient subject is the foundation of all science. But since our sense-perceptions inform us only indirectly of this external world, or Physical Reality, it is only by speculation that it can become comprehensible to us. From this it follows that our conceptions of Physical Reality can never be definitive; we must always be ready to alter them, to alter, that is, the axiomatic basis of physics, in order to take account of the facts of perception with the greatest possible logical completeness. A glance at the development of Physics shows that this axiomatic basis has in fact suffered profound modifications in the course of time.

The greatest alteration in the axiomatic basis of physics—in our conception of the structure of reality—since the foundation of theoretical physics by Newton, originated in the researches of Faraday and Maxwell on electromagnetic phenomena. We shall endeavour, in what follows, to make this more definite by considering also the earlier and the subsequent course of the development of physics.

According to Newton's system, Physical Reality is characterized by the concepts *space*, *time*, *the material particle*, and *force* (interaction between material particles); and physical processes are to be thought of as movements of the material particles in space according to certain laws. The material particle is the sole representative of reality in so far as reality is variable. It was evidently in perceptible bodies that the concept of the material particle had its origin; the particle was conceived on the analogy of the movable bodies, by leaving out their characteristic properties of extension, form, spatial orientation, all their "inner" qualities, keeping only inertia and translation, and adding the concept of force. The material bodies which had thus psychologically been the starting-point in the construction of the concept "material particle" had now themselves to be conceived of as systems of material particles. It is to be observed that this theoretical system is by nature atomistic and mechanical; all activity was to be thought of as purely mechanical—that is, simply as the movement of material particles according to Newton's Laws of Motion.

The most unsatisfactory part of this theoretical system, apart from modern difficulties in the concept of "absolute space", was the theory of light, which Newton, as consistency demanded, also thought of as consisting of material particles. Even at that time the urgency of the question: What happens to the constituent material particles when light is absorbed? must have been acutely felt. Moreover, the introduction into the theory of the two entirely different kinds of material particles that are necessary for the representation of ponderable matter and of light is in itself most unsatisfactory; and later there appear the electric corpuscles as a third kind of particle, with again entirely different properties. It was also a weakness in the foundation of the theory that the forces of interaction, which determine what will happen, had to be assumed as a purely arbitrary hypothesis. Yet a great deal was achieved by this conception of Reality. How, then, did the feeling arise that it must be abandoned?

In order to give his system mathematical form, Newton had first to discover the concept of the differential coefficient, and to enunciate the Laws of Motion in the form of total differential equations —perhaps the greatest intellectual stride that it has ever been granted to any man to make. For this purpose partial differential equations were not necessary, and Newton made no methodical use of them. But they were necessary for the formulation of the mechanics of deformable bodies; this is connected with the fact that in such problems it does not matter in the first place precisely how the bodies are constructed from their material particles.

Thus the partial differential equation first came to theoretical physics as a servant, but by degrees it became its master. This process began in the nineteenth century, when, under pressure of facts of observation, the undulatory theory of light gained acceptance. Light in empty space was conceived to be a vibration of the aether, and it seemed gratuitous to regard this aether as itself a conglomeration of material particles. Here for the first time the partial differential equation appeared as the natural expression of the elementary in physics. In a special branch of theoretical physics the continuous field appeared side by side with the material particle as the representative of

Physical Reality. This dualism, though disturbing to any systematic mind, has to-day not yet disappeared.

If the idea of Physical Reality had now ceased to be purely atomistic, it still remained purely *mechanical*; all change was still to be interpreted as a movement of inertial masses. Indeed, it was not conceived that it could be thought of otherwise. Then came the great revolution for ever linked with the names Faraday, Maxwell and Hertz. The lion's share in this revolution was Maxwell's. He showed that all that was then known of light and electromagnetic phenomena could be represented by his well-known double system of partial differential equations, in which the electric and magnetic fields appear as dependent variables. It is true that Maxwell tried to find a basis or justification for these equations in ideal mechanical constructions, but he used several of these constructions side by side, and took none of them too seriously; it was clear that the equations themselves were all that was essential, and that the field intensities that appeared in them were elementary, not derivable from other simpler entities. At the turn of the century the conception of the electromagnetic field as an irreducible entity was already generally accepted, and the more serious theorists had ceased to believe in the necessity of justifying Maxwell's equations, or in the possibility of providing them with a mechanical basis. Soon, on the contrary, an attempt was made to give a field-theory explanation of material particles and their inertia with the help of Maxwell's theories, but this attempt achieved no ultimate success.

If we leave aside the important *special* results which Maxwell contributed in the course of his life to particular domains of physics, and confine our attention to the modification that he produced in our conception of the nature of Physical Reality, we may say that, before Maxwell, Physical Reality, in so far as it was to represent the processes of nature, was thought of as consisting in material particles, whose variations consist only in movements governed by partial differential equations. Since Maxwell's time, Physical Reality has been thought of as represented by continuous fields, governed by partial differential equations, and not capable of any mechanical interpretation. This change in the conception of Reality is the most profound and the most fruitful that physics has experienced since the time of Newton; but it must be confessed that the complete realization of the programme contained in this idea has so far by no means been attained. The successful physical systems that have been set up since then represent rather a compromise between these two programmes, and it is precisely this character of compromise that stamps them as temporary and logically incomplete, even though in their separate domains they have led to great advances.

Of these, Lorentz's Theory of Electrons must first be mentioned, in which the field and the electric corpuscles appear side by side as complementary elements in the comprehension of reality. There followed the Special and General Theories of Relativity, which, although based entirely on field-theory considerations, have not yet been able to dispense with the independent introduction of material particles and total differential equations. The latest and most successful creation of theoretical physics, namely Quantum Mechanics, is fundamentally different in its principles from the two programmes which we will briefly call Newton's and Maxwell's. For the quantities that appear in its laws make no claim to describe Physical Reality *itself*, but only the *probabilities* for the appearances of a particular physical reality on which our attention is fixed. Dirac, to whom, in my opinion, we owe the most logically perfect presentation of this theory, rightly points out that it appears, for example, to be by no means easy to give a theoretical description of a photon that shall contain within it the reasons that determine whether or not the photon will pass a polarizator set obliquely in its path.

Yet I incline to the belief that physicists will not be permanently satisfied with such an indirect description of Reality, even if the theory can be fitted successfully to the General Relativity postulate. They would then be brought back to the attempt to realize that programme which may suitably be called Maxwell's: the description of Physical Reality by fields which satisfy without singularity a set of partial differential equations.

THE SCIENTIFIC ENVIRONMENT
OF CLERK MAXWELL

BY

Sir Joseph Larmor

SOME apology seems to be due from me for occupying the time of this company: I can plead the insistence of the Cavendish Professor, who perhaps thought that one survival of a past age was required for his scheme of arrangements.

Thus Eddington might now have been telling you how he had abetted Einstein and Minkowski in the construction of a super-intellect, that could take up in a single glance all history, and all locations of the cosmos in all the differing aspects under which it would appear to astronomers of all the ages, who had laboriously built their science from records of exact observations on one of the planets that are rushing through the heavens— thus creating for themselves a transcendental reasoner who would surely be a solipsist philosopher, there being no room for another of his type. In the early days of this hyper-cosmical scheme some of us, after due perplexity, had just given up sharing this ambition as being beyond our range, when Eddington and Davison, abetted by Dyson, came marching back with minute eclipse records from the tropical regions of Africa and Brazil, to tell a war-worn world that the signs were that the cosmos must be so regarded, that by that path one might come somehow to the final understanding of things: and who shall count the number of hours that have since been spent on the alignments of this molluscous chimaera dire? In these latest days Dirac by new symbolic processes, of mentality transcending even those of Hamilton and Grassmann who led a reaction against extraneous spatial frames, continues to instil fresh hope of this achievement of connecting down, into the levels of our concerted human apprehension, these concatenations in the universal outlook of this super-intellect that transcends our space and time, whose fourfold range suggests that there ought to be cognizance also of many other domains of possibility, not coherent but never self-contradictory, that have not dawned upon the restricted perceptions of mankind. We now await the result of Einstein's new efforts, founding on the essential Lagrangian residues of Kantian space: while, by recent reports from the observers on Mount Wilson, the universe, carrying its own space on its back, is expanding headlong into the resuscitated void.

Yet amid the respect that is due to the modernist schemes of transcendental relativity, we would not forget here the astronomical relativity of just two centuries ago (1732): how a modest country parson, a friend of Newton's later years who had helped him in exploring orbits of comets for the *Principia* and so resolving their mystery, had shown how to disentangle the rational system of the heavens out of the welter of recorded observations, of all astronomers wherever and however travelling in their orbital paths—thus promptly providing the necessary supplement to Newton's guiding theory, by originating an exact science of practical astronomy, and leaving it in such security for all subsequent ages that not even an Einstein could suggest amendment of its scheme except in a few excessively minute and still possibly uncertain details.

One might also be enjoying here the advantage of listening to Niels Bohr, who, if he were not a Scandinavian, might almost be claimed to be a Cambridge man. For has he not in student days here drunk in the same academic air that Clerk

Maxwell had breathed, imbibed the same partiality for mechanical or astronomical models, however provisional, of the under-structure of the world, hopeful and undismayed by their patent imperfections.[1] When by a simple arithmetical computation he showed us here how the universal constant of Planck, then coming into its full and totally unexplained recognition, could be connected up through the spectroscopic generalizations with the very small number of the other absolutes of physical science, did we not wonder what possibly he could make out of the bare coincidence, even after his case had been fortified by an opportune prediction about the spectrum of helium in relation to that of hydrogen? Yet we know that he has gone on feeling his way into new outlooks in spectroscopic fundamentals, and new elucidations of the scheme of groupings of the chemical elements previously so laboriously essayed, all gained perhaps by refusing to become submerged in the attractions of auxiliary algebra and so keeping the whole rapidly expanding field of phenomena fluently within a single direct outlook, almost

[1] So too had Maxwell constructed provisional models of an aether, to be afterwards transcended, while Kelvin parted company by persisting in their necessity.

emulating the manner of the super-intellect created by Einstein and Eddington. They say that Clerk Maxwell was a poor lecturer, perhaps because, like Lagrange, he could not hold his active mind fixed down to the small detail that he ought to be expounding, but allowed it to wander after the hints of new views that could not help constantly opening out before him. So perhaps with our friend Bohr; he might want to instruct us about the correlations of too many things at once, out of the repertory floating before his vision, so that as in the case of Maxwell it might on the whole be easier to study his systematic pronouncements on the printed page.

The welcome presence of Zeeman will recall to us how the discovery of a very minute phenomenon, which in its simpler aspects was quite reasonable and almost obvious, could, by exploration of intricate developments that it presented, due largely to his own lifelong experimental investigation, reverberate throughout the furnaces of the universe, under the inspiring lead of G. E. Hale, as a primary clue to fundamental knowledge.

Then again, to pass to the most modern stage of this marvellous progress—but not more wonderful

as we shall maintain than that of a century ago—
you might be listening to G. P. Thomson, another
Cambridge man in the very strictest sense of that
epithet, who in his experimental quest has even
had in startling fashion to endow the electrons
with pulsating wings, originating a new and delicate technique; and thereby appears to be also
driven towards cultivating, in concert with L. de
Broglie, the companionship of that same super-
intellect who fuses the presentations of all observers
into his one scheme, as a stage in the reconstruction
of a new physical world from the *débris* of the one
of which he has been illustrating the insufficiency.
And there are de Broglie and Schrödinger who
independently have not shirked to push on abstract
speculation "and follow knowledge, like a sinking
star, beyond the utmost bounds of human thought",
who have thereby brought back to science a very
remarkable, if as yet hardly coherent, reward.

Then there is also the recent weighty treatise
of R. H. Fowler, exhibiting enviable analytical
power and width of thought, which by a mode
of metaphor recently familiar may be taken to
afford an indication that the originator of the
universe of which it treats must have been a
statistician.

And one need hardly refer to Rutherford himself, who, contemporary here with Townsend and McLennan, a quarter of a century ago summarily shattered by his experimental deductions the implications of the historical atomic theory, and has been paying the recognized penalty ever since in superintending the painful construction of a new scheme of things nearer to the heart's desire. This problem has moreover now become alluring with far greater precision than even did that of Kepler, in the hands of Aston, who, following up and improving on his master J. J. Thomson, has been digging deep into the primordial roots of chemical structure, revealing astonishing simplicities of which many of us here can hardly expect to survive to see the necessary elucidation. Nor need one mention here the new practical science of ultra-optics, developed and cultivated by W. L. and W. H. Bragg, with great experimental mastery by Siegbahn, and also in new directions by Debye and his associates, which shows up the internal arrangements of crystals and even the architecture of the more regular molecules; which again has imposed on its originators the vast task of patiently reconstructing crystallography, opening out developments, theoretical and technical, in that science of which there appears to be no end. And there are the thermionic phenomena, now becoming precise yet merely sporadic even a few years ago, cultivated assiduously by Richardson assisted by the vacuum technique, but in the early unfathomed stage already switched off to absolutely magical technical account by the practical acumen of Fleming and his successors.

One recalls here the recent loss of Michelson and of Lorentz, whose far-reaching activities of genius, experimental and constructive, went back in time far enough to link up intimately with those of Maxwell.

All these fields of activity hark back to a purely intellectual lead, flowing largely from the genius of Clerk Maxwell, and to the experimental impulse that his ideas at length acquired from the observations of Heinrich Hertz, who knew how to recognize the electric waves that were all around him, because of one advantage over other searchers, that as a theorist he was aware of what he was to expect and could fit an interpretation to casual observations such as encouraged further pursuit.

As regards none of these topics do I venture, in modernist days—perhaps in part being sensible of confusions due to attempting to combine interests in unduly wide ranges of contemplation—to try to hold your attention. Some day this tantalizing wealth of unexpected and conflicting clues to physico-chemical origins will fuse together, and a widened synthesis will stand out in clear light before the vision of mankind.

It is proposed instead now to pass under brief review a topic not so remote from present activities as may appear; namely the history of some of the perplexities that stood in the way of the original settlement of the basic and still largely indispensable concepts of historical physical science, especially a group of them which, if now apparently rudimentary, must yet still rank among the profound reaches of knowledge. Man is greater than his finished thought, and his processes are equally worthy of study. If modern development rests on the mental activity of Maxwell, as of W. Thomson and Helmholtz and the other giants, in their turn Maxwell and Thomson had interacted with Faraday and were his interpreters, just as he for long years studied and improved upon Ampère according to his lights. It was as early as 1821, a year after Oersted and Ampère, that the electromotor came into being: it presented itself unobtrusively in the simple and beautiful form of Faraday's so-called sustained electromagnetic rotations. The mere question why it did not occur to anyone that it must be a reversible engine, and so could generate a current by working backward from power supplied, illustrates the limitations that could arise from the undeveloped state at that time of physical concepts now the most elementary; though Carnot, with his supreme emphasis on cyclic reversible operation, was to arrive within three years in those fecund days. The personal record of Faraday's activity, carefully prepared from the sources by himself for vol. II of his *Experimental Researches*, gives an idea of the continued effort during ten years, including a serious mental collapse for four of them, before he could demonstrate the actual induced currents at a date (1831) which has recently been commemorated. The rapidity with which there had followed, after the preliminary very subtle exploration by Ampère, the presumed electrotonic state of the intervening medium, its presumed state of stress, to be developed later into arresting mathematical precision by Maxwell, with culmination in 1831 in the law of induction, is surely the proof of Faraday's penetrating familiarity during all those ten years with his relevant new ideas of physical interaction to which the want of mathematics perhaps fortunately had drawn him.

But if it is easy now to recognize that Clerk Maxwell was a main pioneer, in part as a disciple in Thomson's school, from whom largely the vast ranges of progressive knowledge revealed in the modern world of experimental electrics and optics have had their origin, there is another domain of thought, even more fundamental and far more delicate, on whose progress he has left an ineffaceable mark, though this time rather as one of an army of more or less tentative collaborators. He has here to be considered in relation to the giants who guided the evolution of an abstract universal science of energy, Carnot, Joule, Mayer, Waterston, W. Thomson, Clausius, Helmholtz, Boltzmann, Willard Gibbs, to whom may be added the engineers Rankine and Clapeyron and J. Thomson, also Kirchhoff and Rayleigh; all resting ultimately on the French experimental school who discovered the basic thermal properties of gases and provided the material for the constitutive ideas of Avogadro and Ampère and in fact of modern molecular chemistry. The modes in

which trains of ideas arise in the human mind, and
often independently in different minds when the
times are ripe, ought surely to be instructive and
worthy of close study as a discipline in method:
and the more permanently unsettled and difficult
are the subjects to which these ideas of the masters
pertain, the greater ought to be the profit thus
derivable, especially when lapse of time may be
held to have cleared away personal and local pre-
ferences and controversies that can clog contemporary activities but in the scientific fraternity
are not presumed to survive. Perhaps in no domain of science has the neglect of history been
more conspicuous than in physics, and this applies
especially to the most elusive because most universal chapters, of which we propose to treat.

But before passing on, we may advert to the
fascination for Maxwell of the Hamiltonian reconstruction of dynamics, now in its centenary years
—not as expanded by Jacobi into an abstract
theory of the field of differential equations, but as
a connection through Lagrangian variations between distant relations of a group of states of the
same system through a characteristic function
defining relations across a distance, whose suffi-
ciency Hamilton was content to base on practical

applications. Maxwell's beautiful developments in optical systems illustrated this interest: and also showed up his expertness in symmetric geometrico-algebraic analysis of physical activities, which seems to have come easy to him, but for which it is surprising that he found so much time, even as a relaxation and perhaps in the solitudes of Galloway, during his brief career.[1]

In the famous synthesis of Lagrange, the *Mécanique Analytique*, dynamical science had assumed this form of an abstract discipline such as could be developed strictly by itself, and then compared with the phenomena of the world, which had suggested it, to see how far its artificial scheme ran parallel. Its foundations could be amended so as to improve upon this correspondence: they could even be extended into dimensions beyond what

[1] In hydrodynamics Maxwell also had affinity with a powerful school of vortex theorists, including after the lead of Kelvin the names of J. J. Thomson, H. Lamb, W. M. Hicks, Love, Bryan following on the pioneers Helmholtz and Kirchhoff. The settling down of vortices originating in slip at a sharp trailing edge—after the more recent ideas of Lanchester following on Helmholtz and Kelvin—into a steady state of aerial motion with circulation round the air-wing combined with a succession of vortex rings thrown off at its two extremities, is now familiar in aeronautic technology.

could be directly confronted with experience, or even visualized by human faculty, thus opening out the germs of new sciences, leading in Maxwell's phrase "to conquests new in worlds not yet created". For example the scheme of relative motion is developed by Lagrange from abstractions of algebra and compared with phenomena: thus initiating the importance of invariance relative to the frame of reference, which passed on to the Helmholtz-Lie theory of groups of displacements. "Les méthodes qui j'y expose ne demandent ni raisonnements géométriques ni méchaniques, mais seulement des opérations algébriques, assujéties à une marche régulière et uniforme."

Maxwell was perhaps the earliest, in the great consolidating electric memoir of 1865, at any rate after Hamilton who had been impelled by the urge of his analogies from optical rays and images thirty years earlier, to apply the algebraic dynamical scheme formulated by Lagrange to the exploration of phenomena whose relations were in part unrevealed: more systematic exposition, pointing to a foundation on varying action, came some years later (1867) by Thomson and Tait, with the insistence on the illustration by the Cotes basic optical theorem of apparent distance which can dominate all simple relations in that field of rays, or orbits, and with startling applications to hydrodynamics, which under more rigorous formulation became the kernel of its application in navigation and aeronautic practice.

A science of thermodynamics leapt into being fully formed, in 1824, in a fashion too novel and strange in relation to current trends of ideas to gain recognition at the time, coming from the brain of perhaps the supreme scientific genius of the last century, Sadi Carnot, who in his short detached career may remind one of Pascal, but by the practical vein appropriate to his service as military engineer and also present in Pascal as well as Galileo and Torricelli, was very different from the more abstract intellects, also of unfulfilled renown, contemporary with him, such as Abel and Galois. But the vast subject instinctively mapped out by him as regards its essential ideas could not become a progressive science until there was a basic theory of matter, on which the primary crude conception of the nature of heat, otherwise unfathomable in any exact sense, could take form in some definite way; and that arrived with sufficient precision only through the formulation of the kinetic theory of gases. Who was it that made reasonably secure this foundation? Here even more than in most constructive interpretations of nature, the course of development has been ever interesting, a veritable "progress through illusion to the truth".[1] An examination of even the standard sources still reveals great variety of plausible but often irreconcilable approaches to the subject. The entangled history of this domain has hardly yet been critically explored: and it may not be out of place to advert —it may be my last effective opportunity—to the conviction of a lifetime regarding a scientific development, never free from disputation, in which the thoughts of Maxwell played an essential part in the later formative stages. The tentative recent expansions, basing themselves on new types of phenomena opened out by the far wider range and power of modern atomic experimenting, do not invalidate the historical science of thermodynamics: they limit its range to the quite coherent domain in which these phenomena, important

[1] This was the arresting title of a set of Hulsean Lectures preached before the University in the writer's undergraduate days by Dr E. A. Abbott, eminent through a long but isolated career as theologian, philologist and educator. If memory serves, Professor Maxwell was regular in attendance.

mainly at very high and very low temperatures, so mostly outside direct human experience, are of negligible account. Yet it is hard to evade the conclusion, from detailed comparison of the many authoritative expositions, that the final authentic mode of wider correlation has not yet been attained.

[The remaining part of this essay, largely an investigation into the historical origins of thermodynamics and the kinetic theory, from Carnot and Joule onwards, in relation to Maxwell's position therein, has developed to too great length and is not suitable for the present volume: it is proposed to publish it later.]

CLERK MAXWELL'S METHOD

BY

Sir James Jeans

CAMPBELL and Garnett open their life of Maxwell by describing him as: "One who has enriched the inheritance left by Newton and has consolidated the work of Faraday—one who impelled the mind of Cambridge to a fresh course of real investigation".

On this account, they say, he "has clearly earned his place in human memory". Maxwell's reputation has increased with the passage of time, and we of to-day should state his claim to fame in far higher terms. He has no doubt earned a place in our memories through following worthily in the footsteps of great predecessors, but he has earned a higher place on account of his breaking of new scientific ground. In his hands electricity first became a mathematically exact science, and the same might be said of other large parts of physics. He was also conspicuous in his generation for looking beyond the superficial appearance of phenomena—if not to the underlying reality, at least to the realization that the reality lay very deep indeed.

His serious scientific work commenced in the middle years of the last century, a time when science was advancing at a greater pace than had ever before been known, except possibly for two brief periods of a few years each, in the lives of Galileo and Newton respectively.

The science of that day had its own clearly-marked characteristics. Its main concern was to discover the facts of nature with a view to turning them to immediate, utilitarian ends; it was the science of the Industrial Revolution, which had its apotheosis in the Great Exhibition of 1851. Michael Faraday, whom we have just been honouring, embodied the scientific spirit of that age, and we can best see the special quality of Maxwell's contributions to science by viewing them against Faraday as a background.

Faraday was above all things an experimenter; Tyndall wrote of him: "A good experiment would make him almost dance with delight". His scientific range was immense, yet he was primarily a chemist and remained a chemist at heart throughout his life. By contrast Maxwell began his scientific career by studying the geometry of the

regular solids, and by reading papers on plane
curves to the Royal Society of Edinburgh while
still a boy. Both these interests were of a purely
abstract nature. He amused himself, it is true,
with scientific toys, and took a certain amount of
experimental litter up to Cambridge with him. At
the time of writing his first mathematical papers,
he was making copper seals with the device of a
beetle, and discovering with surprise that the body
of a dead beetle was not a good conductor of
electricity. In his more mature years, we find him
noting with interest that the skin of a dead cat is
electropositive to that of a live dog, and wondering
what sort of reaction there would be if both animals
had been alive when their skins were rubbed together. Yet I do not think he ever felt any deep
interest in experimental methods, nor showed any
outstanding skill as an experimenter. When experiments were not an amusement to him, they were
merely fodder for mathematical investigation; he
was content that others should perform them, and
reap the harvest which often accrues without un-
due delay to the successful experimenter.

As a consequence, Maxwell did not "alter the
face of civilization" as Faraday did, or at least
did not alter it so immediately or in a manner so

obvious to the eye. It could hardly have been written of him during his lifetime, as it was of Faraday, that "our daily life is full of resources which are the results of his labours; we may see at every turn some proof of the great grasp of his imaginative intellect".

Faraday had used his clear vision and consummate skill as an experimenter to explore those strata of nature which lie immediately under our hands; Maxwell used his clear vision and consummate skill as a theorist to explore the deeper strata in which the phenomena of the upper strata have their origin.

Faraday discovered the fact of nature; Maxwell, taking the known fact, endowed it with a theoretical setting, and with a precision such as enabled its utility to be extended almost indefinitely. Just because this kind of work lies deep down, a long time must often elapse before its utility first becomes apparent to the world. If an example is needed, one is to hand in Maxwell's electromagnetic theory, and the electromagnetic waves which he saw were implied in the theory. Nearly a quarter of a century was to pass before these waves were detected in the laboratory, and yet another quarter century before they took their place in everyday life. While they exemplify the time lag which must inevitably intervene between seed-time and harvest, between the research in pure science and its utilitarian application, they provide an even more outstanding illustration of how great the value may be—even if often a deferred value—of research in pure science carried out for the sake of pure science alone, and with no motive other than to understand the innermost workings of nature.

Mathematicians fall into two more or less distinct types. One regards mathematics mainly as a manipulation of symbols; if these are handled without mistake according to the orthodox rules, the correct answer is bound to come out at the end. To this school the intervening steps have no practical significance—the transformation of formulae and solutions of equations are not associated with physical ideas; they are mere mathematical steps to a mathematical end.

The mathematical physicist of to-day must perforce belong to this school; most of the symbols he uses convey no physical meaning to his mind; he can explain and predict the whole course of atomic nature in terms of the behaviour of a symbol—the ψ of Schrödinger's wave-mechanics—but he cannot tell us what ψ means in physics; and I for one doubt if he will ever be able to do so.

To the second school every mathematical step in the analysis is full of physical significance; every symbol is alive, and represents a clear physical concept in the mind of the mathematician. This is true even when the symbol does not correspond to anything existing in nature. An obvious instance is the magnetic vector-potential, which Maxwell had pictured as the measure of the intensity of Faraday's "electrotonic state"; he described the components of this vector-potential as "Electrotonic Functions" or "Components of Electrotonic Intensity".

Maxwell belonged naturally and whole-heartedly to this second school. From the very beginning his mathematical ideas were not only guided but controlled by a strong sense of physical reality. His Cambridge coach Hopkins, after describing him as "unquestionably the most extraordinary man I have met in the whole range of my experience", went on to say, "it appears impossible for Maxwell to think incorrectly on physical subjects; in his analysis, however, he is far more deficient". Perhaps the most conspicuous instance of this quality of his mind is to be found in his first contribution to the Dynamical Theory of Gases, a subject which Maxwell was later to make peculiarly his own. The idea that the properties of matter could be explained in terms of the rapid motions of its ultimate invisible particles was of course at least as old as Lucretius. In more modern times various physicists, including Bernoulli, Clausius and Joule, had shown that many of the properties of gases could be explained by regarding the gas as composed of showers of swiftly moving particles. They had, however, pictured all the particles as moving with the same speed. Maxwell not only saw, what was of course in any case obvious, that the speed of a particle was necessarily changed each time it collided with another particle; he also saw that further progress of an exact kind could only be made by allowing for the different speeds of the various particles. Thus, it became a problem of absolutely fundamental importance to find a mathematical formula for the speeds of the different molecules which make up a gas. Maxwell, by a train of argument which seems to bear no relation at all to molecules, or to the dynamics of their movements, or to logic, or even to ordinary common sense, reached a formula which, according to all precedents and all the rules of scientific philosophy, ought to have been hopelessly wrong. In actual fact it was subsequently shown to be exactly right, and is known as Maxwell's law to this day.

It was this power of profound physical intuition, coupled with adequate, although not outstanding, mathematical technique, that lay at the basis of Maxwell's greatness. Yet he was perhaps less remarkable in the possession of a vivid physical imagination than in the strict control he kept over it. He never allowed it to run away with him. No matter how clearly he saw physical concepts in his mind's eye, he never made the mistake of identifying them with ultimate physical reality. He saw too deeply into things ever to imagine that what he saw was the ultimate stratum of all—final and absolute truth. The average scientist of the 'seventies would, I suppose, have unhesitatingly insisted that light must ultimately be of the nature of undulations in an ethereal medium. Now hear Maxwell's own words:

> The changes of direction which light undergoes in passing from one medium to another are identical with the deviations of the path of a particle in moving through a narrow space in which intense forces act. This analogy was long believed to be the true explanation of the refraction of light; and we still

find it useful in the solution of certain problems, in which we employ it without danger as an artificial method. The other analogy, between light and the vibrations of an elastic medium, extends much farther, but, though its importance and fruitfulness cannot be over-estimated, we must recollect that it is founded only on a resemblance *in form* between the laws of light and those of vibrations.

It sounds almost like an extract from a lecture on modern wave-mechanics—and a very good lecture, too.

By never identifying his physical pictures with reality, he left himself free to discard one picture and adopt another as often as expediency or convenience demanded. He described his method of procedure in the following words:

If we adopt a physical hypothesis, we see the phenomena only through a medium, and are liable to that blindness to facts and rashness in assumption which a partial explanation encourages. We must therefore discover some method of investigation which allows the mind at every step to lay hold of a clear physical conception, without being committed to any theory founded on the physical science from which that conception is borrowed.

Faraday had tried to picture electrostatic forces as stresses running along lines of force in an elastic medium; Maxwell, wishing to make Faraday's ideas amenable to exact mathematical treatment,

7-2

developed what we now describe as the "Hydro-dynamical analogy", justifying his procedure as follows:

By referring everything to the purely geometrical idea of the motion of an imaginary fluid, I hope to attain generality and precision, and to avoid the dangers arising from a premature theory professing to explain the cause of the phenomena. If the results of mere speculation which I have collected are found to be of any use to experimental philosophers, in arranging and interpreting their results, they will have served their purpose, and a mature theory, in which physical facts will be physically explained, will be formed by those who by interrogating Nature herself can obtain the only true solution of the questions which the mathematical theory suggests.

It was the great misfortune of nineteenth-century science that this philosophic spirit deserted it, just at about the time of Maxwell's death. Its close saw great scientists averring that the one thing they were sure of was "the reality and substantiality of the luminiferous ether", and in all seriousness calculating that its density must be millions of times that of lead. Maxwell was able to construct the most impossible pictures of phenomena, learn all there was to be learned from them, and then discard them immediately, before they could hamper his further progress; his successors, having constructed equally impossible pictures, proceeded solemnly to announce them to a wondering world as scientific realities.

At this time the only science which could be treated with complete mathematical exactness, at any rate according to the general scientific opinion of the day, was astronomy. Newton's law of gravitation was believed to be exact, and the sun and planets could be treated as particles to an almost perfect approximation. It ought accordingly to have been possible, at least in theory, to predict the motions of the planets with almost complete accuracy.

In the physical sciences, on the other hand, it had been necessary to introduce simplifying assumptions, which might be good or bad, but were never under any circumstances believed to be perfect. As a consequence, the mathematical treatment of problems of physics was at best an approximation, and often enough a quite bad approximation. A conspicuous instance was provided by the mathematical theory of elastic solids, which Maxwell himself tried to improve, although with only indifferent success.

When Maxwell turned to the electromagnetic theory of light, he pictured it in terms of a medium whose properties could be specified completely in

terms of a single mathematical constant. He saw
that if the value of this constant could once be
discovered, it ought to become possible to predict
all the phenomena of optical theory with complete
mathematical precision. Maxwell showed that the
constant in question ought to be merely the ratio
of the electromagnetic and electrostatic units of
electricity, and his first calculation suggested that
this was in actual fact equal to the constant of the
medium which measured the velocity of light.

The first mention of the great discovery comes
in a letter which he wrote to Michael Faraday
under the date 19th October, 1861:

> I suppose the elasticity of a sphere to react on the electrical
> matter surrounding it, and press it downwards. From the
> determination by Kohlrausch and Weber of the numerical
> relation between the static and magnetic effects of electricity,
> I have determined the *elasticity* of the medium in air, and
> assuming that it is the same with the luminiferous ether, I have
> determined the velocity of propagation of transverse vibra-
> tions.
>
> The result is 193,088 miles per second. Fizeau has deter-
> mined the velocity of light as 193,118 miles per second by
> direct experiment.

The situation was comparable in its dramatic
intensity to the great moment when Newton first
tested his law of universal gravitation, by calcu-

lations on the distance of the moon. By a piece of bad luck, Newton used an inaccurate value for the earth's diameter, and this led to such poor numerical agreement that he put his theory aside for almost twenty years. Maxwell had the opposite kind of luck; the two numbers quoted above agreed to within thirty miles a second, although oddly enough both are in error by more than 6000 miles a second. When Maxwell came to publish his paper, "A Dynamical Theory of the Electromagnetic Field", probably the most important and far-reaching paper he ever wrote, he gave the two velocities in terms of kilometres a second, and it became apparent that the numbers were in nothing like such good agreement as he had originally supposed. The ratio of the units as determined by Kohlrausch and Weber (310,740,000 metres a second) agrees almost exactly with the value given by Maxwell in his letter, but Fizeau's velocity of light (314,858,000 metres a second) is nowhere near to the figure he had quoted. Happily he seems to have realized that the velocity of light was not at all accurately known, and so did not allow himself to be deterred, as Newton had been, by a substantial numerical disagreement.

After his electromagnetic theory of light, Max-

well's most important and best known work is
that on the Kinetic Theory of Gases. Here again
he started with a perfectly definite picture—a vast
number of small, hard and perfectly elastic spheres,
acting on one another only during impact; again
he was careful to explain that his model was only
provisional—to be discarded if the results deduced
from it failed to agree with experiment. At first
it was brilliantly successful, and Maxwell was able
to explain the viscosity and diffusion of gases and
their conduction of heat in terms of the free paths
of colliding molecules. For reasons we now understand, it failed, as it was bound to fail, before the
problem of specific heats.

Maxwell states his conclusion in the very
guarded words:

> We have now followed the mathematical theory of the
> collisions of hard elastic particles through various cases, in
> which there seems to be an analogy with the phenomena of
> gases....We proved that a system of such particles could not
> possibly satisfy the known relations between the two specific
> heats of all gases.

Next year he read a paper "On Bernoulli's
Theory of Gases" to the Oxford Meeting of the
British Association. After saying he had travelled
through the region of the clashing of molecules,

he ended his paper with the following significant words:

But who will lead me into that still more hidden and dimmer region where Thought weds Fact—where the mental operation of the mathematician and the physical action of the molecules are seen in their true relation? Does not the way to it pass through the very den of the metaphysician, strewed with the remains of former explorers and abhorred by every man of science?

The list of discrepancies between theory and observation was not yet complete. Maxwell had found that if the molecules of a gas were hard elastic spheres, the viscosity would be proportional to the square-root of the absolute temperature. He performed some experiments on actual gases, which he described in his Bakerian Lecture delivered before the Royal Society in 1866. He found that a gas imitated his model in becoming more viscous as it was heated, but the rate of increase was greater than was to be expected if the gas were a crowd of hard elastic spheres. Maxwell believed he had found that the viscosity of an actual gas was exactly proportional to the absolute temperature, and saw that this would be the property of a gas in which two molecules repelled one another according to the inverse fifth power of their distance apart. He accordingly set to work to study mathematically the properties of such a gas. By a happy chance this particular law proved to be peculiarly suited to exact mathematical investigation, and the problem was fully worked out in the great memoir *On the Dynamical Theory of Gases*, which appeared late in 1866. In this we probably see Maxwell's astounding powers of physical intuition and his very considerable mathematical capacity at their highest. Unhappily he was still vulnerable in the Achilles' heel to which his Cambridge coach had drawn attention thirteen years before—elementary analysis and simple arithmetic. It was a very human failing, which many of us will welcome as a bond of union between ourselves and a really great mathematician, but it had disastrous results. Maxwell's belief that the viscosity of an actual gas varied directly as the absolute temperature proves to have been based on faulty arithmetic, and the conclusions he drew from his belief were vitiated by faulty algebra. Thus the great investigation stands rather as a model of method than as a monument of achievement. Yet possibly the result it reached was as good as any other that could have been reached at the time. For a third of a century had still to elapse before the quantum theory came into being, and it alone, as we now know, could have provided the key to the mysteries which Maxwell was attempting to solve.

It was not in keeping with Maxwell's methods that he should finish off a piece of work so completely that nothing could be added to it; his plan was rather to open up wide vistas which would provide work in their detailed exploration for whole generations yet to come. It was certainly so with the Dynamical Theory of Gases, with the Electromagnetic Theory of Light, and with his electrical work in general. Yet there is one exception to this general statement. His first serious mathematical investigation was that on the stability of Saturn's rings, which gained for him the Adams Prize in the year 1857. In this, he showed that Saturn's rings could not be continuous in structure, but must consist of a crowd of small particles, myriads of miniature moons of the great planet. Here we have almost the only example of a piece of work which Maxwell himself completed and left in perfect form. His conclusions received observational confirmation 38 years later when the American astronomer Keeler obtained spectroscopic proof that the outermost portions of the rings were rotating less rapidly than the inner portions.

Only the man of science himself ever knows how little freedom he has in choosing his scientific path. One problem leads him inevitably on to another, and it may well be that a man's first scientific research contains within it the germs of a whole life's work. Although there is no definite evidence, many think that Maxwell's study of the particles of Saturn's rings led him directly and inevitably into the realm of the kinetic theory of gases, in which so much of his life was spent. However this may be, when he crossed the bridge from Astronomy to Physics he left behind him for ever the prospect of becoming a great mathematical astronomer—but only to become the greatest mathematical physicist the world had seen since Newton.

MAXWELL'S LABORATORY

BY

William Garnett

Magna opera Domini exquisita in omnes voluntates ejus.

THIS quotation from the Vulgate, carved in bold letters on the great oak doors of the Cavendish Laboratory, perfectly expresses Maxwell's mental attitude to the researches which he hoped would be carried out by himself and his students in the laboratory which he had designed. For though the drawings were ably prepared and the builder's work supervised by Mr W. M. Fawcett, M.A., of Jesus College, the Architect appointed by the University, the Laboratory and its equipment as originally designed sprang from the mind of Maxwell as complete as Athene from the brain of Zeus.

A magnetic room occupied the whole width of the building at the east end of the ground floor and was heated by means of copper pipes. The tables were stone slabs four feet square supported on brick piers, which stood on separate foundations and were carried up through the floor without touching it. One of these tables was occupied by

the great electrodynamometer of the British Association which, with other instruments used in the
determination of the British Association Standards
and with the original Ohm, was deposited in the
Laboratory by the Association. The second table
was occupied by a unifilar magnetometer of the
Kew pattern. High up in the north wall was a
window through which observations could be taken
for the determination of the Meridian. The next
room was the clock room provided with a stone
pier for the support of the principal clock and a
stone frame for the suspension of an experimental
pendulum, both on foundations independent of the
floor. This frame was subsequently used by Poynting for the repetition of "the Cavendish Experiment" in a modified form.

Adjoining the clock room was the balance room
with a north light and then followed the heat room
in which was placed Maxwell's apparatus for determining the viscosity of air, which formed the
subject of the Bakerian Lecture in 1866. In the
battery room was provided a Daniell's battery of
the type devised by Sir William Thomson with
trays about 2 feet square lined with lead, in which
the zinc plates were supported on 1 inch porcelain
cubes, the resistance of each cell being 0·16 ohm.

The small workshop which completed the ground floor may be said to have led to the establishment of the Engineering Laboratory and of the Cambridge Instrument Company.

The working tables throughout the whole building were made independent of the floors. Openings were left in the floors and beams were supported on brackets on the walls of the rooms below. On these beams blocks were fixed which were finished flush with the floor without touching it, and on these blocks the table legs rested. Throughout the Laboratory also, stone window sills were provided to each window, both inside and out, finished to the same level, so that instruments might be placed partly inside and partly outside the windows. In the west wall a small window was inserted on the second floor opposite the passage for a heliostat, from which a beam could be sent uninterruptedly for the whole length, 120 feet, of the Laboratory. A number of apertures were made in the first and second floors for long suspension wires reaching from the roof to the ground floor.

The first floor comprised the general laboratory, 60 feet by 30 feet, the Professor's private room, with the quadrant electrometer on a shelf in one corner, the apparatus room fitted with glass show cases containing much historical apparatus, the preparation room and the lecture theatre accommodating 180 students.

On the second floor were a room for experiments in acoustics, a drawing office, a room for radiant heat and two rooms for optical experiments, besides a dark room with blackened walls. Over the apparatus room was a large room for experiments in high tension electricity. This room was provided with a special apparatus for drying the air and a window communicated with the lecture theatre 17 feet above the floor of the theatre. Through this window leads could be taken to the lecture table to provide high tension electricity if the air in the lecture room was too damp for its generation by glass machines in that room. A space was left in the landings and floors at one corner of the staircase tower for a Bunsen water pump and a mercury gauge more than 50 feet in height.

If we enquire into the purposes for which Maxwell anticipated that the Laboratory should be used, it is better to let Maxwell speak for himself.

In a letter to Lord Rayleigh dated 15th March 1871, Maxwell wrote:

If we succeed too well, and corrupt the minds of youth till they observe vibrations and deflections and become Senior

Optimes instead of Wranglers, we may bring the whole University and all the parents about our ears.

But in his public utterances he expressed no fear of the laboratory exciting the opposition of either the dons or the parents. In his inaugural lecture in October 1871 he distinguished between "experiments of illustration" and "experiments of research". Of the former he said:

The aim is to present some phenomenon to the senses of the student in such a way that he may associate it with the appropriate scientific idea. When he has grasped this idea the experiment which illustrates it has served its purpose.

Of the latter:

In experimental researches, strictly so called, the ultimate object is to measure something which we have already seen— to obtain a numerical estimate of some magnitude.

Experiments of this class are the proper work of a Physical Laboratory. In every experiment we have first to make our senses familiar with the phenomenon; but we must not stop here, we must find which of its features are capable of measurement, and what measurements are required in order to make a complete specification of the phenomenon. We must then make these measurements and deduce from them the result which we require to find.

This characteristic of modern experiments—that they consist principally of measurement—is so prominent that the opinion seems to have got abroad that in a few years all the great physical constants will have been approximately estimated, and that the only occupation which will then be left to men of science will be to carry on these measurements to another place of decimals.

If this is really the state of things to which we are approaching, our laboratory may perhaps be celebrated as a place of conscientious labour and consummate skill, but it will be out of place in the University, and ought rather to be classed with the other great workshops of our country, where *equal ability* is directed to more useful ends.

But we have no right to think thus of the unsearchable riches of creation, or of the untried fertility of those fresh minds into which these riches will continue to be poured.... The history of Science shews that even during that phase of her progress in which she devotes herself to improving the accuracy of the numerical measurement of quantities with which she has long been familiar, she is preparing the materials for the subjugation of new regions, which would have remained unknown if she had been contented with the rough methods of her early pioneers. I might bring forward instances gathered from every branch of Science, shewing how the labour of careful measurement has been rewarded by the discovery of new fields of research, and by the development of new scientific ideas.

* * * * * *

Our principal work in the laboratory must be to acquaint ourselves with all kinds of scientific methods, to compare them and to estimate their value. It will, I think, be a result worthy of our University, and more likely to be accomplished here than in any private laboratory, if, by the full and free discussion of the relative value of different scientific procedures, we succeed in forming a school of scientific criticism, and in assisting the development of the doctrine of methods.

True to the principles he had enunciated, Maxwell did not attempt to attract to the Laboratory, or to his own lectures, large classes of undergraduates. His idea was to get a band of graduates each doing a piece of work after a short course of training in measurements. During the first two or three years the Laboratory was attended by a small group of Fellows and Post-graduates (including Pirie, W. D. Niven, Chrystal, Hicks, Fleming, Glazebrook, Shaw, Sedley Taylor, Poynting, MacAlister, Schuster and J. E. H. Gordon). But Maxwell by no means despised the "experiment of illustration" and it was not long before he encouraged his demonstrator to deliver courses of lectures to candidates for the Natural Sciences Tripos and to medical students on mechanics and physics, and to illustrate them by apparatus constructed in the Laboratory workshop and fit for no other useful purpose.

SOME MEMORIES

OF

PROFESSOR JAMES CLERK MAXWELL

BY

Sir Ambrose Fleming

As one of the small class of two or perhaps three University students who attended Maxwell's last Course of University Lectures and also worked in the Cavendish Laboratory during the sessions 1877–78 and 1878–79, the last two years of his life, I am gratified to have the privilege of joining in these recollections of a teacher whose supreme genius left its mark on every subject he investigated or taught.

I came up to Cambridge in October 1877 chiefly with the object of benefiting by his lectures and laboratory teaching and of putting myself in a position to understand a little better his electrical investigations and writings.

His great treatise on Electricity and Magnetism, first published in 1873, had shown all those who had only an experimental acquaintance with the subject, derived from such old-fashioned books as de la Rive's treatise, that there had arisen a new mode of treating it based on the translation and extension of Faraday's ideas into mathematical language, and we all desired to have instruction in it.

I thought that could best be obtained by coming to the fountain-head and trying to learn something of it from Maxwell himself.

I had obtained an open entrance Science Exhibition at St John's College in the early part of 1877 and after surmounting the barrier of the Little-go, I began to attend Maxwell's lectures in the beginning of the Lent term in 1878. I well remember my surprise at finding a teacher who was everywhere regarded as the greatest living authority on his subjects lecturing to a class of two or three students in place of the 100 or more attentive listeners he would have had in any Scottish or German university. The reason for this was better understood later on.

Yet it was characteristic of Professor Maxwell that he lectured to the two or three students with the same care and completeness of treatment that he would have done if his lecture room had been filled to its utmost capacity.

In the last year of his life the only two attendants at his lectures were an American gentleman,

Mr Middleton, who was a resident at St John's, and myself.

Maxwell lectured us on Thermodynamics in the October term and Electricity in the Lent and May terms.

I took careful notes and wrote them out with some additions after each lecture, and these bound volumes of lectures, which have been in my possession for 52 years, are now by kind permission of Lord Rutherford to be donated to the Cavendish Laboratory as memorials of Maxwell's University lectures in those two final years of his life.

In the Electricity Course he gave us a new and powerful method of dealing with problems in networks of linear conductors. Kirchhoff's corollaries of Ohm's law had provided a means only applicable in the case of simple problems in which one could foresee the direction of flow of current in each conductor. But that was not possible in complicated networks.

Maxwell initiated a new method by considering the actual current in each wire to be the difference of two imaginary currents circulating in the same direction round each mesh of the network. In this way the difficulty of foreseeing the direction of the real currents was eliminated.

The solution of the problem was then reduced to the solution of a set of linear equations and the current in any wire could be expressed as the quotient of two determinants.

After Maxwell's death in 1885 I communicated a paper to the Physical Society of London in which this method of Maxwell's was extended so as to give an expression for the electrical resistance of any network between any two points.

In October 1878 I asked Professor Maxwell to advise me as to the practical work I should do in the Cavendish Laboratory. He suggested the comparison of the resistance coils which were made for the British Association Committee and were intended to represent a resistance of 1 ohm.

These seven or eight coils were made with wires of different alloys and had marked on them the temperatures at which they were intended to have a resistance of exactly 1 ohm. As a matter of fact they had all slightly altered and the problem presented was to determine which coil and at what temperature was the most probable B.A. unit of resistance. The investigation consisted in determining the difference in the resistance of any coil placed in water at a certain known temperature and that of one particular coil kept at 0° C. The differences were taken by a method suggested by Professor Carey Foster which had previously been employed at the Cavendish Laboratory by Professor Chrystal.

As the ordinary bridge was not convenient for this purpose, I devised a special form of resistance balance which from its resemblance in shape to a well-known musical instrument Maxwell used to call "Fleming's banjo".

The results of the work were plotted in a series of curves on a chart from which the most probable value of the mean B.A. unit could be determined in terms of a particular coil set at a certain temperature. The absolute value of this resistance was later on re-determined by the late Lord Rayleigh, Sir Arthur Schuster and Sir Richard Glazebrook, and it was found that the absolute value was not 1 theoretical ohm or 10^9 c.g.s. units of resistance but about 1·5 per cent. too small or about 0·985 ohm.

At about that time in 1876 Alexander Graham Bell had invented and made known in England his speaking telephone and T. A. Edison had patented in 1877 his "carbon button" transmitter. In 1878 D. E. Hughes had invented the microphone and shown to electricians the remarkably

useful qualities of a "loose contact" between
conductors, a thing which electricians had always
previously disliked and avoided as much as possible.

In May 1878 Maxwell gave a lecture in the
Senate House on the telephone. It was a brilliant
discourse illustrated by flashes of wit, apt analogies
and much learning but it was not of the type most
useful to convey to unscientific hearers an idea of
the mode in which a telephone operates.

Maxwell in short had too much learning and
too much originality to be at his best in elementary
teaching. For those however who could follow
him his teaching was a delight.

He had an immense knowledge of the history
of science and could illustrate the simplest facts
by some telling anecdote or phrase. Also in all his
experimental work he chose the most effective and
simple means for achieving his result.

During the time I worked in the Cavendish
Laboratory he confined himself to giving his lectures and coming in perhaps for an hour in the
afternoon and chatting with those who were doing
any work there.

He was occupied chiefly at that time in com-
pleting the work of editing and seeing through the
press the original scientific papers of the Honour-

able Henry Cavendish which had been placed in his hands by the then Duke of Devonshire. Cavendish amongst other things had made fairly accurate measurements of the dielectric constants of some insulators such as glass, wax, resin, etc., and also the relative electric conductivities of iron and of water.

Cavendish did this by the crude method of taking electric shocks through his body and adjusting the circuits in some way so as to make the "feel" of the shocks of about equal intensity. Maxwell repeated carefully all the principal experiments of Cavendish to ascertain how far such comparisons were possible.

Cavendish had a servant called Richard who was occasionally called in to take shocks and give his opinion about them, although we are not told how far he enjoyed these supplementary duties.

In the same way Maxwell invited several of us who were working in the laboratory to assist in a similar manner. I remember rather vaguely helping him one afternoon in this manner but I cannot remember at this distance of time how far I proved to be an accurate "shockmeter".

There were however great differences in this respect between observers, and Maxwell measured at one time the electrical resistance of several men with the Wheatstone Bridge taken from hand to hand and found, as might have been expected, that the "boating men" had very high resistance owing to the state of their hands.

It was always a matter of surprise to me and others that Maxwell never seems to have attempted to obtain any experimental proof of the existence of the electromagnetic waves, which he first predicted theoretically in his paper on "A Dynamical Theory of the Electromagnetic Field" published in the *Transactions* of the Royal Society of London for 1865, a paper which may be considered to be one of the greatest productions of the human mind.

It was not however until eight years after Maxwell's lamented death that H. Hertz succeeded in creating experimentally these electromagnetic waves in space and in showing how to detect them.

G. F. FitzGerald had previously suggested that they might perhaps be created by the oscillatory discharge of a Leyden jar. Sir Oliver Lodge had probably come very near to it in his researches on this subject; and D. E. Hughes, even so far back as 1878, had without doubt created them, though without clear recognition; and had detected them by a metallic filings coherer. Probably the reason that Maxwell did not turn his attention to the matter was in the absorption of his time and attention first in writing his great treatise, next in the engagement of his time in the design and organization of the Cavendish Laboratory, and thirdly in the time taken up in editing the Cavendish papers, which appears to have occupied most of the last five years of his life.

During the last term in May 1879 Maxwell's health evidently began to fail, but he continued to give his lectures up to the end of the term and one of the notebooks I am presenting to the Cavendish Laboratory contains the notes of the last lecture he gave in his life.

When he had finished that lecture he had, unconsciously perhaps to himself, and certainly to the two students who formed his audience, performed the last duty he was ever to conduct in the University of which he was so great an ornament.

To have enjoyed even a brief personal acquaintance with Professor Maxwell and the privilege of his oral instruction was in itself a liberal education, nay more, it was an inspiration, because everything he said or did carried the unmistakable mark of a genius which compelled not only the highest admiration but the greatest reverence as well.

CLERK MAXWELL AND WIRELESS TELEGRAPHY

BY

Sir Oliver Lodge

In 1865 Clerk Maxwell was attempting to assimilate the ideas of Faraday about the interrelation of the electric and magnetic fields as typified by Faraday's conception of lines of force. He perceived that a magnetic field was wrapped round a current, and that a current could equally well be wrapped round a magnetic field, that in fact the relation between them was reciprocal, and could be expressed mathematically by what he subsequently called "curl". When the two fields coexisted in space, the reaction between them could be expressed as the curl of a curl, and this he found simplified down to the well-known *wave* equation, the velocity of propagation of the wave being the reciprocal of the geometric mean of an electric and a magnetic constant. This velocity for a time he called the number of electrostatic units in a magnetic unit, and proceeded to devise experiments whereby it could be measured. Experiments made by him at King's College, London, resulted in some near approach to the velocity of light; so that thenceforth, in his mind, light became an electromagnetic phenomenon.

He was assisted in his ideas by an imaginative constructive model of the ether, a model with rolling wheels and sliding particles, which subsequently he dropped, presumably as being too complex for reality, and remained satisfied with his more abstract equations, which were reproduced in his great treatise in 1873. The publication of this treatise was heralded and announced to the world by the pure mathematician Professor H. J. S. Smith of Oxford, in his Presidential Address to Section A of the British Association, then meeting in Bradford. The following is a quotation from this address:

> In the course of the present year a treatise on electricity has been published by Professor Maxwell, giving a complete account of the mathematical theory of that science, as we owe it to the labours of a long series of distinguished men, beginning with Coulomb, and ending with our own contemporaries, including Professor Maxwell himself. No mathematician can turn over the pages of these volumes without very speedily convincing himself that they contain the first outlines (and something more than the first outlines) of a theory which has already added largely to the methods and resources of pure mathematics, and which may one day render to that abstract

science services no less than those which it owes to astronomy. For electricity now, like astronomy of old, has placed before the mathematician an entirely new set of questions, requiring the creation of entirely new methods for their solution.

I was present at the meeting, and soon afterwards conceived the idea of making the production and demonstration of these electromagnetic waves the work of my life. In 1878, at Dublin, I made the acquaintance of G. F. FitzGerald, and together we discussed many proposals or plans for producing and detecting them. In 1883 FitzGerald communicated two papers to Section A, in one of which he calculated (on Maxwell's principles) the energy lost by radiation, from an alternating circular current of moment $\pi a^2 i_0 = M$ and of period T, as

$$\frac{8\mu\pi^4}{3T^4c^3} M^2, \quad \text{or} \quad 10^{-29} M^2 N^4,$$

where N is the frequency; while the other paper was "On a Method of producing Electro-magnetic Disturbances of comparatively Short Wave-lengths", and is abstracted in the annual volume of the British Association in the following words:

This is by utilising the alternating currents produced when an accumulator is discharged through a small resistance. It would be possible to produce waves of as little as 10 metres wave-length, or even less.

Henceforth it may be said that the remaining difficulty was not the production of the waves, but their detection and demonstration.

By 1887 and 1888 they had been definitely produced and detected by Hertz, his method of detection being the small sparks which to his surprise were produced in conductors in the neighbourhood of an electric discharge. Hertz's great discovery was announced by FitzGerald in his Presidential Address to Section A at Bath in 1888. At the same time I exhibited the waves running along and being guided by wires.

By 1889 I had hit on the coherer method of detecting electric surgings, and demonstrated the existence of waves in various forms, among others being the sparking of the wall-paper at the Royal Institution during Leyden jar discharges on the table.

By 1894 Branly's improved method with a tube of iron filings was employed for demonstration of all their optical analogies; and in August of that year at Oxford it was shown that telegraphic messages across space could be sent in that way. In 1896 Mr Marconi came over to this country with the beginnings of a practical arrangement for this kind of signalling. He aroused the interest of Sir William Preece, and thereafter a system of wire-

less telegraphy began its early struggling but afterwards triumphant career.

That is a sketch as nearly accurate as I can make
it of the way in which Maxwell's theoretical investigation led through Hertz's discovery to actual
practice on the large scale.

In 1897 I showed how tuning could be used for
increasing the sensitiveness of response and for
selection between different stations; and in 1901
onwards Marconi found that the waves would curl
round the earth to America and beyond. In 1904
Fleming discovered that a vacuum tube was a more
efficient detector than any coherer or crystal, and
began the application of it to telephony as well as
telegraphy; in 1908 De Forest improved the valve,
making magnification possible. And so gradually,
by many inventions, a world-wide system of wireless transmission was inaugurated, until to-day
a wireless set is a domestic apparatus in nearly
every household.

Thus the whole system of wireless telegraphy is
a development of the original and surprising theory
of Clerk Maxwell, embodying in mathematical
form the experimental researches of Faraday; while
the rapid and world-wide extension of the system
must be credited to the energy and perseverance
of Senatore Marconi and his co-workers.

EARLY DAYS AT THE
CAVENDISH LABORATORY

BY

Sir R. T. Glazebrook

CLERK MAXWELL returned to Cambridge as Cavendish Professor in 1871, but for some years previously he had been in close contact with the Mathematical teaching in the University.

In 1850 there had been a Royal Commission, and the Commissioners, after calling attention to the manner in which Physical Optics was studied, had stated that no reason could be assigned why other great branches of Natural Science should not equally become objects of attention, or why Cambridge should not become a great school of physical and experimental, as it was already of mathematical and classical, instruction.

The Tripos regulations had been in force, practically unaltered, since 1848, and though Natural Philosophy was mentioned in the schedule the interpretation given to the phrase was limited. It did not include Heat, Electricity, the theory of Elastic Solids, or any but the quite elementary parts of Hydrodynamics. These subjects were however included in a new scheme approved in 1868 which was to come into force in 1873. But the University had no means of teaching the experimental side of Physics and a Syndicate was appointed in 1868 to make proposals to remedy the defect. Maxwell examined in the Tripos in the four years 1866, 1867, 1869 and 1870 and his work as Examiner did much to show the Syndicate the need for systematic instruction, especially in Heat, Electricity and Magnetism. The recommendation of the Syndicate that a Professorship should be established was accepted by the Senate and culminated in Maxwell's appointment. The generous offer of the Duke of Devonshire, the Chancellor, in 1870, to provide a Laboratory had made it possible for the University to house the Professor in a fitting manner, and the Cavendish Laboratory was opened with due ceremony in June 1874.

My own life as an undergraduate began in October 1872, a few months before the first examination (January 1873) under the new schedule, including Heat, Electricity, and Magnetism. Maxwell was additional Examiner, a fifth Examiner additional to the two Moderators and two Examiners, appointed with a view to the introduction

of the new scheme. He was followed by Thomson
and Tait and in 1876, the year I took my degree,
by Rayleigh.

It was a time of great interest; there were few
or no text-books, Fleeming Jenkins' *Electricity and
Magnetism* in Longmans' Text-book of Science
Series appeared in 1873, Maxwell's *Heat* in the
same series was published about the same time
as well as his *Electricity and Magnetism*. The subjects were new to our teachers as well as to ourselves. Thomas Dale of Trinity was my coach. I
still have the notes on Thermodynamics in which
he boiled down to a consistency suitable for undergraduate digestions Thomson's work on Thermodynamics and the then to us mysterious second
law, or his papers on Electrostatics. Fortunately
for us, Lamb, then a young Lecturer in the College,
had grasped the greatness of Helmholtz's essay
on Vortex Motion and Thomson's writings on
Hydrodynamics. His lectures, afterwards to form
his book on Hydrodynamics, were a revelation to
those of us who had only read Besant's *Hydrodynamics*. My copy of Maxwell's *Electricity* was
bought, I think, during the Long Vacation of 1874.
I have it still, a few Chapters or Sections of Volume I
marked with Dale's R, others that we should now

think most interesting with an 0. Volume II I hardly touched until after the Tripos when, in the summer of 1876, I went with Dale to Buttermere and we read much of it together.

Meanwhile the Professor was establishing himself, the Laboratory was being built; lectures were given, and some students, rather senior to myself, had started work. Of this I knew nothing at the time.

After the Tripos I was told I ought to read for a Trinity Fellowship. The Dissertation system had just been introduced. Hitherto Fellowships had been awarded on the result of an examination, and Dale suggested I should go to the Cavendish to learn some practical physics and see if I could find a subject suitable for a dissertation. And so a number of us went, among them Poynting and Shaw.

The first morning Maxwell explained to us, standing round a table in the big Laboratory, Wheatstone's Bridge method of measuring resistance. I regretted that my first visit had not been a few weeks earlier for in one of his questions Rayleigh had asked us to "explain the Wheatstone's Bridge method of measuring resistances" and had set an easy rider on its use. Unfortunately such practical matters as the measurement of resistance had not been included in my superficial study of Maxwell and I could touch neither question nor rider.

Soon we were measuring resistances and learning to set up and adjust various pieces of electrical apparatus. To fix and adjust the mirror in a Thomson's galvanometer was not quite easy for clumsy fingers. There were no regular demonstrations—the organization of practical classes for the Natural Sciences Tripos came later in Lord Rayleigh's day. The Professor gave us some experiment to carry out and the necessary apparatus; we were left much to ourselves, William Garnett of St John's was there as Demonstrator to assist or to suggest means of getting out of a difficulty.

After a short time I was set a task which I think any student with only a few weeks' Laboratory experience would consider alarming. There had been no rigorous experimental proof that Ohm's law was true, and Chrystal, afterwards Professor at Edinburgh, and Saunders, who had taken their degrees in 1875, were testing it, using a method devised by Maxwell who was a member of a Committee of the British Association appointed in 1874 to investigate experimentally its accuracy. Doubts had been raised by a paper submitted by Schuster.

The method involved the measurement of a resistance, first when carrying a large current and afterwards when the current was greatly reduced. They were using as their source of electromotive force a number of sawdust tray Daniell's cells and desired to know whether in the course of the work serious variations took place in these.

I was set to measure periodically the E.M.F. of these, using a Thomson's Quadrant Electrometer of the large pattern. Few now know the instrument and no one would think of employing it for the purpose. To maintain the needle at constant potential a small trap-door electrometer and a replenisher were attached, and it was necessary to set this accurately before taking a measurement. To make the thing work was a trial, perhaps a useful one. Chrystal's result, as stated in his Report to the British Association in 1876, was that the resistance of a conductor, which for an infinitely small current is 1 ohm, is not altered by 10^{-12} when carrying a current of 1 ampere.

The British Association resistance coils had been brought by Maxwell to the Cavendish and in the course of their work were carefully intercompared by Chrystal and Saunders.

At this time the Professor was engaged in editing

the Electrical papers of the Hon. Henry Cavendish, placed at his disposal for the purpose by the Chancellor, and this led to his designing an improved form of the experiment of the sphere and hemispheres for verifying the law of the inverse force; the experiment was carried out by Donald Mac-Alister and proved that if the force follows the law r^{-2+q}, then q must be between $\pm$ 1/21600.

Schuster came to Cambridge in 1876 and carried out some spectroscopic investigations; others working were Heycock and Clayden; Shaw who had gone to Berlin to work under Helmholtz was recalled in 1879 to undertake an investigation for the Meteorological Council. Poynting after his election to a Fellowship at Trinity in 1878 had come back to continue his determination of the mean density of the Earth. In 1878 Fleming was at work on the comparison of the British Association standards of resistance, and designed his bridge for the more accurate and ready comparison of two nearly equal resistances. This was exhibited at the recent Guthrie Lecture to the Physical Society.

But to return to my own experiences. The measurement of the E.M.F. of a battery or some work on the properties of a stratified dielectric which I was set to do did not seem likely to afford material for a Fellowship Dissertation. I was interested in Physical Optics and on reading Stokes' Report to the British Association on Double Refraction, was attracted by a suggestion it contained for testing Fresnel's theory of the form of the Wave Surface. Stokes had applied the method to a uniaxial crystal and on enquiry if he still thought that it was of importance to examine a biaxial crystal in the same way I received every encouragement from him. The goniometer he had used was available, and he lent it to me with some of his apparatus. Schuster obtained for me a crystal of aragonite and had it cut in accordance with Stokes' suggestions in a manner which allowed the examination of two considerable arcs of the wave surface. The work was the more interesting for Lord Rayleigh had recently suggested a modified theory of double refraction according to which the inertia in a crystal is a function of the direction, the elasticity being constant. On Fresnel's theory the inertia is the same in all directions while the elasticity differs.

Maxwell it is true had published his *Electricity and Magnetism* which contains the outline of the Electromagnetic theory of light, but at the time optical problems were discussed in terms of some form of Elastic Solid theory. The Electromagnetic theory as is well known leads to Fresnel's wave surface.

Maxwell readily promised me facilities for the work and I was installed in a gloomy cell with black walls and very insufficient ventilation, the first room on the top floor, then fitted for optical work, where a powerful sodium flame made the atmosphere far from pleasant. Garnett had installed a few tools in the basement of the Laboratory but there was no regular mechanic. We had to put together ourselves such minor fittings as we needed. However, I profited by lessons received from the village carpenter when a boy and was very pleased when I designed and made a small clamp which would hold my crystal in any desired position. As the work progressed, Maxwell looked in from time to time. I remember on one occasion when I was in some difficulty I asked his advice. In reply he said that others in the Laboratory had been asking questions which formed a good thick crust around his brain; mine would take time to soak in. But in a day or two he came in again saying if you do so and so I think you will find it is all right. In due course the work was finished and the Dissertation written. I was able to show that over the two arcs examined, the measured velocity was very closely that given by Fresnel's theory. In only two directions was there a difference of as much as 0·0004 in the refractive index, and a second set of observations made it clear that most of this difference was due to difficulties in making the observations. In general the differences were much less. The conclusion reached was that except possibly in the neighbourhood of the optic axis, where observation was difficult, the experiments confirmed Fresnel's theory. In the case of Lord Rayleigh's theory differences of some five times the above amount were found; it did not account for the facts.

The Dissertation was referred to Maxwell and on his report I was elected a Fellow. "We could not resist", the Master said when I interviewed him after the admission, "the strong case Professor Maxwell put before us in your favour." I happened to meet Maxwell at Bletchley Station on the evening of the election—we were both returning from the north—and remember clearly his kindly greeting and his cordial congratulations.

With the original crystal it was only possible to examine two arcs of no very great length, and it was thought desirable to extend the work by measurements on a second crystal. This occupied the best part of another year with much the same result and the whole[1] was communicated to the Royal Society by Maxwell in June 1878.

There appeared to be some evidence that the small variations from Fresnel's theory observed might depend on the wave-length, and by the advice of Stokes and Maxwell it was thought desirable to investigate this. The work was thus continued using prisms of Iceland spar instead of aragonite. For this there were two reasons: larger prisms could be obtained giving more light, and Stokes' own few experiments on Iceland spar had never been published. The experiments were repeated using the hydrogen spectrum as a source of light, with the result that "it appears that Huyghens' construction represents the result of experiment for the three rays of the hydrogen spectrum to a degree of approximation comparable with the probable error of the experiments". The average difference between theory and experiment in the refractive index of the line C was less than 0·00007.

Maxwell communicated the paper[2] containing

[1] "On Plane Waves in a Biaxial Crystal", *Phil. Trans.* Part I, 1879, pp. 287–375.

[2] "Double Refraction and Dispersion in Iceland Spar", *Phil. Trans.* Part II, 1879.

these results to the Royal Society in June 1879.
It was the last piece of work in which I had his
help and advice. He was then ill, and left Cambridge in June for his Scottish home at Glenlair
hoping to recover; but he gradually grew worse,
and I never saw him again. He died at Cambridge
on November 5, 1879.

The above is a very incomplete account of the
Cavendish in Maxwell's days; it is rather the story
of how he helped and encouraged one young
student in his desire to advance Natural Knowledge, and perhaps it could hardly be otherwise;
for it is a sample of the method adopted towards
all those who worked under him.

We were trained to be independent workers;
encouraged to find some problem we might hope to
solve, helped in our difficulties, and guided to success. We realized, I think, the greatness of our
leader, admired, though perhaps without a full
appreciation, his own contributions to knowledge,
and loved him for himself. It was a privilege to
work with him for three short years; it is a privilege
now to contribute in any way to his memory.

BY

Sir Horace Lamb

My own recollections of Maxwell relate only to the early days of his professorship at Cambridge. A number of young mathematical students, either candidates approaching the Tripos, or recent graduates, had been keenly interested in the proposals for the foundation of a Chair of Experimental Physics. The study of Natural Philosophy, in a certain limited sense, had of course long flourished in the University, but this was a new departure, and there was naturally much speculation as to the choice of the new Professor. It had been rumoured that Sir W. Thomson had been invited, and even that Helmholtz had been approached. When Maxwell was ultimately appointed, in 1871, he was, I am afraid, little more than a name to many of us, except that on one or two occasions he had been responsible for some highly original questions set in the Mathematical Tripos. It is to be remembered, in our excuse, that at that date neither of his two important

treatises had been published, and that the study
of current scientific literature by mathematical
students, so far as it was indulged in, was looked
upon as an eccentricity likely to damage them
in the keen struggle for places. At any rate we
looked forward with interest and some enthusiasm to the advent of the new professor, and took
care not to miss his inaugural lecture. The circumstances attending the delivery of this were rather
unusual. There had been only a rather casual
announcement, which escaped the knowledge of
all the leading personalities in the University, and
when the time came the lecture was delivered, not
in the Senate House as might have been expected,
but in an out of the way lecture room, and to a
score or so of students.

The sequel was rather amusing. When, a few
days later, it had been announced with proper
formality, that Professor Maxwell would begin his
lectures on Heat at a certain time and place, the
dii majores of the University, thinking that this was
the first public appearance, attended in full force,
out of compliment to the new Professor, and
it was amusing to see the great mathematicians
and philosophers of the place such as Adams,
Cayley, Stokes, seated in the front row, while

Maxwell, with a perceptible twinkle in his eye, gravely expounded to them the relation between the Fahrenheit and Centigrade scales, and the principle of the air-thermometer.

It was rumoured afterwards, and it is not incredible, that Maxwell was not altogether innocent in this matter, and that his personal modesty, together with a certain propensity to mischief, had suggested this way of avoiding a more formal introduction to his Cambridge career.

The subsequent lectures on Heat, and later ones on Electricity, followed very much the lines of his books, which were then on the stocks. They had a great interest and charm for some of us, not so much for the sake of the subject matter, which was elementary, as in the illuminating glimpses we got of the lecturer's own way of looking at things, his constant recourse to fundamentals, and even his expedients when in a difficulty, the humorous and unpremeditated digressions, the occasional satirical remarks, and often a literary or even poetical allusion. There was nothing forced or frivolous about this; the interpolated remarks and illustrations were always spontaneous and apt. Something of their quality may be gathered from the occasional essays on various topics which are to be found in his collected works. It was interesting also to hear his appreciations, sometimes restrained, but more often generous, of the great mathematicians of the past. I especially remember the way in which he would refer to Ampère, and Gauss, and Stokes.

He had his full share of misfortunes with the blackboard, and one gathered the impression, which is confirmed I think by the study of his writings, that though he had a firm grasp of essentials, and could formulate great mathematical conceptions, he was not very expert in the details of minute calculation. His physical instincts saved him from really vital errors.

The effect of his lectures was such that when his great book on Electricity and Magnetism, long looked forward to, appeared in 1873, there was a great rush to the booksellers for copies. It was studied eagerly, and formed, as is well known, the starting point for a long series of investigations by his pupils. But here I am getting beyond the topic of merely personal recollections.

To a more intimate knowledge of Maxwell, beyond an occasional visit to his house, I can make no claim. He had two toys which he would sometimes bring out to entertain fresh visitors.

One was the "dynamical top", intended to illustrate various points in the classical theory of rotation. It is, I think, to the paper which he wrote on this apparatus that one can trace the modern revival of interest in the variation of latitude. The other toy was a form of ophthalmoscope which he had independently invented, probably in connection with his researches in Dioptrics. He was wont to demonstrate the use of this on himself and his friends, including his dog, which he had trained to become a patient and accommodating "subject".

Milton Keynes UK
Ingram Content Group UK Ltd.
UKHW041520181024
449640UK00009B/80